Understanding Evo-Devo

Why do the best-known examples of evolutionary change involve the alteration of one kind of animal into another very similar one, like the evolution of a bigger beak in a bird? Wouldn't it be much more interesting to understand how beaks originated? Most people would agree, but until recently we didn't know much about such origins. That is now changing, with the growth of the interdisciplinary field of evo-devo, which deals with the relationship between how embryos develop in the short term and how they (and the adults they grow into) evolve in the long term. One of the key questions is: Can the origins of structures such as beaks, eyes, and shells be explained within a Darwinian framework? The answer seems to be yes, but only by expanding that framework. This book discusses the required expansion, and the current state of play regarding our understanding of evolutionary and developmental origins.

Wallace Arthur was one of the founders of the interdisciplinary field of evo-devo (evolutionary developmental biology) in the 1980s. He was a founding editor of the journal *Evolution & Development*. His interests are focused on the evolutionary origins of novel types of animal, and how these origins may differ from 'routine evolution'.

The ***Understanding Life*** series is for anyone wanting an engaging and concise way into a key biological topic. Offering a multidisciplinary perspective, these accessible guides address common misconceptions and misunderstandings in a thoughtful way to help stimulate debate and encourage a more in-depth understanding. Written by leading thinkers in each field, these books are for anyone wanting an expert overview that will enable clearer thinking on each topic.

Series Editor: Kostas Kampourakis http://kampourakis.com

Published titles:

Understanding Evolution	Kostas Kampourakis	9781108746083
Understanding Coronavirus	Raul Rabadan	9781108826716
Understanding Development	Alessandro Minelli	9781108799232
Understanding Evo-Devo	Wallace Arthur	9781108819466

Forthcoming:

Understanding Genes	Kostas Kampourakis	9781108812825
Understanding DNA Ancestry	Sheldon Krimsky	9781108816038
Understanding Intelligence	Ken Richardson	9781108940368
Understanding Metaphors in the Life Sciences	Andrew S. Reynolds	9781108940498
Understanding Creationism	Glenn Branch	9781108927505
Understanding Species	John S. Wilkins	9781108987196
Understanding the Nature–Nurture Debate	Eric Turkheimer	9781108958165
Understanding How Science Explains the World	Kevin McCain	9781108995504
Understanding Cancer	Robin Hesketh	9781009005999
Understanding Forensic DNA	Suzanne Bell and John Butler	9781009044011
Understanding Race	Rob DeSalle and Ian Tattersall	9781009055581
Understanding Fertility	Gab Kovacs	9781009054164

Understanding Evo-Devo

WALLACE ARTHUR
National University of Ireland, Galway

CAMBRIDGE
UNIVERSITY PRESS

University Printing House, Cambridge CB2 8BS, United Kingdom

One Liberty Plaza, 20th Floor, New York, NY 10006, USA

477 Williamstown Road, Port Melbourne, VIC 3207, Australia

314–321, 3rd Floor, Plot 3, Splendor Forum, Jasola District Centre, New Delhi – 110025, India

79 Anson Road, #06–04/06, Singapore 079906

Cambridge University Press is part of the University of Cambridge.

It furthers the University's mission by disseminating knowledge in the pursuit of education, learning, and research at the highest international levels of excellence.

www.cambridge.org
Information on this title: www.cambridge.org/9781108836937
DOI: 10.1017/9781108873130

First published 2021

Printed in the United Kingdom by TJ Books Limited, Padstow Cornwall

A catalogue record for this publication is available from the British Library.

ISBN 978-1-108-83693-7 Hardback
ISBN 978-1-108-81946-6 Paperback

'Wallace Arthur treats his readers to an eminently readable but still deeply rooted introduction into one of the most significant achievements of evolutionary biology: how evolutionary developmental biology put the organism back into the centre of evolutionary thinking.'

Günter P. Wagner, Yale University, USA

'Evo-devo deals with the multiple connections that exist between the biological processes of evolution and development. However, as an interface subject, there is a plurality of views on its content and its boundaries. In spite of that, Wallace Arthur has succeeded in writing an extremely clear and highly accessible guide to this fascinating, multifaceted discipline. Using the concept of "developmental repatterning" as a common thread, the book provides a balanced view of evo-devo, covering its main achievements and future challenges. This is an ideal entry point for the non-specialist, but also a stimulating read for the practitioner who wants to consider her/his research in a wider perspective.'

Giuseppe Fusco, University of Padova, Italy

'Occasionally I feel that the field of Evolution and Development has lost its way, becoming submerged in myriad examples and details that don't expand our understanding of life. Wallace's book expounds the intellectual underpinnings of Evolution and Development, leads us through the key questions, and finally shows how the details and examples inform our future understanding. This book provides not just a guide to Evolution and Development, but also a spur to refocus and redouble our efforts to use development to help understand the evolution of life on Earth.'

Peter Dearden, University of Otago, New Zealand

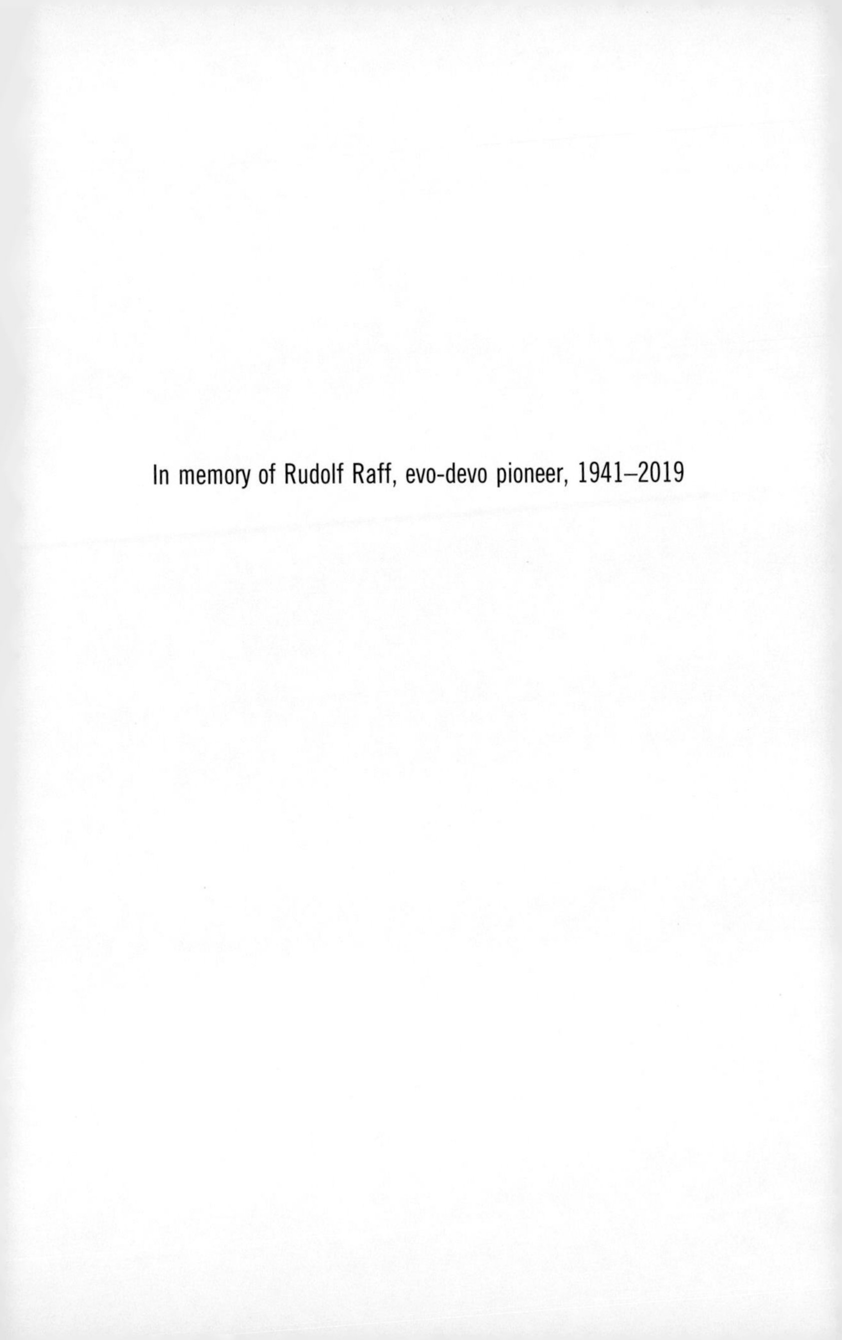

In memory of Rudolf Raff, evo-devo pioneer, 1941–2019

Contents

Foreword

Understanding Evo-Devo, by Wallace Arthur, takes readers on a fascinating voyage across the animal kingdom, and beyond. The author shows that, contrary to some representations, evolution and development are not two distinct processes but, rather, two entirely intertwined ones. Evo-devo, which is short for evolutionary developmental biology, studies evolution and development, and especially their interactions. In particular, it studies both the evolution of development – that is, how developmental processes evolve – and the developmental basis of evolution – that is, how development structures the evolution of the features of organisms. Wallace Arthur has produced a concise and informative book, written with passion and clarity, which provides readers with a coherent understanding of how evolution occurs through changes in development, and of how the study of developmental processes reveals that distantly related organisms with very different body forms have deeper similarities than you might previously have thought. When you come to realize why and how the same genes are crucial for the developmental processes of morphologically very different organisms, you will likely be astonished by the unity of life. Also, evolution cannot be studied by looking at populations and genes alone; rather, individuals and their developmental trajectories must be taken into account too, as this book convincingly shows. This is a must-read for anyone interested in understanding life.

Kostas Kampourakis, Series Editor

Preface

About 40 years ago, a new branch of science was born. Its nickname is evo-devo, and it is generally known by this, since its full name, evolutionary developmental biology, is a bit cumbersome. Its aim is to understand the ways in which the short-term process of biological development (including embryogenesis) is intertwined with the long-term process of evolution. Like any new branch of science, it has roots that extend back into the past – at least into the nineteenth century in this case. But the evo-devo of today is a far cry from the activities that characterized its deepest roots. Today both development and evolution can be explained in terms of genes – their interactions with each other and with various other players ranging from proteins and cells to environmental variables such as temperature and population density. A common language enables us to link evolution and development in a way that was impossible before the 1980s.

The aim of this book is to describe evo-devo, its historical roots, and its interactions with the flanking disciplines of evolutionary and developmental biology, in a way that is accessible to non-specialists. The anticipated readership thus includes students in all areas of the biological sciences, professional biologists who specialize in other fields, and the 'educated layperson'. I have tried to keep this final category of reader firmly in mind while writing. I have assumed a high-school level of biological knowledge but no more. Thus I expect readers to know what the following things are, at least in general terms: DNA, proteins, cells, embryos, larvae, life cycles, inheritance, and natural selection. However, I do not expect readers to have come across more specifically evo-devo entities and concepts, and so I will explain these where they first arise, for example: developmental bias, evolutionary novelties, heterochrony, body plans, homeobox, toolkit genes, the evo-devo hourglass, and gene co-option.

A word about the figures. There are 20 of these in the book, spread through all the chapters. I have kept them as non-technical as possible. Some of them are only referred to in the chapter in which they occur, while others are referred to intermittently throughout the book. Of the latter, four are especially important as ways of locating things against a broad genetic, taxonomic, and palaeontological background. These four, and their locations in the book, are as follows: the broad structure of the living world (Figure 1.1, page 3); within this, the structure of the animal kingdom (Figure 7.1, page 116); a classification of types of genes, including the developmentally important toolkit genes (Figure 4.2, page 71); and the geological timescale (Figure 7.2, page 118).

Now, a word about the references. A book of this kind should not be cluttered with superscripts, footnotes, or other technical or overly prescriptive ways of linking the text to other sources of information. So what I have done is as follows. When there is a connection between an author mentioned in the text and a source listed at the end of the book, I give the year as well as the author's name in the relevant piece of text, though not in a fixed style. Thus wherever you see a name and a date that are reasonably close together, there is a source to look up at the end of the book if you feel like doing so. On the other hand, if there is a name and only a vague timespan, e.g. 'the 1980s' or 'the early twentieth century', then there is no corresponding entry in the reference list. References are grouped by chapter at the end of the book.

Finally, a word about the use of italics, which is more important than you might think. In evo-devo, there are two standard uses of italics, one that comes from taxonomy, and one from genetics. Regarding the taxonomic use, we always italicize the name of a species and the genus to which it belongs, as in *Homo sapiens* or our close relative *Gorilla gorilla* (the latter showing that on rare occasions the genus and species names are the same as each other). In contrast, we do not italicize the names of any of the higher-level taxa to which the species concerned belong – in this case, for example, Mammalia and Vertebrata. Regarding the genetic use, we always italicize gene names but not the names of the corresponding proteins, so that the two may be readily distinguished. For example, the gene *hedgehog*, which we will meet in the first chapter, makes a protein called Hedgehog. This gene and protein were first discovered in flies, but are found widely in many sorts of animals, including humans and, as it turns out, hedgehogs. The widespread occurrence of the same developmental genes across very different-looking animals is a key discovery of evo-devo. For more information on this and other discoveries, read on . . .

Acknowledgements

Many people kindly agreed to read draft material – anything from a single chapter to the whole manuscript. Those who are subject specialists kept me on my toes regarding accuracy, and those who are not did likewise regarding accessibility. I am very grateful to everyone who helped in these complementary respects, and whose composite efforts thus considerably improved the book, namely: Michael Akam, Sean Carroll, Andy Cherrill, Ariel Chipman, Michael Coates, Katrina Halliday, Ronald Jenner, Kostas Kampourakis, Chris Klingenberg, Colin Lawton, and Sandro Minelli. I would also like to thank my son, Stephen Arthur, who produced all the final artwork and made a major contribution to the cover design.

My interactions with Cambridge University Press have been a pleasure, as always, and I would particularly like to thank Katrina Halliday for her invitation to write this book for the Understanding Life series. I was ably helped along the way by Olivia Boult, and Sam Fearnley oversaw a very smooth path through the production process. I thank Chloe Bradley in advance for her help on the marketing side of things; I have every confidence that she will be as outstanding in that domain as she was with my previous book. Hugh Brazier improved the clarity of countless bits of writing at the copy-editing stage; this is the third book we have worked on together, and by now I feel that we have become e-friends as well as e-colleagues. Finally, I would like to thank the series editor, Kostas Kampourakis, for putting up with my robust defence of those sections of text that I felt didn't needed to be changed, and for his kind comments about the book in the Foreword.

1 What is Evo-Devo and Why is it Important?

What is Evo-Devo?

The two great creative processes of biology are evolution and development. You and I, as adult human beings, are products of both. Evolution took about four billion years to make the first human from a unicellular organism that emerged from the primordial soup. Development, in the form of embryogenesis together with its post-embryonic counterpart, takes less than 20 years to produce an adult human from a different unicellular organism – a fertilized egg or zygote. By this measure, development operates more than 200 million times faster than evolution. However, despite their very different timescales, the two great creative processes of biology are intrinsically interwoven. Evo-devo is the scientific study of this interweaving. Its full name is evolutionary developmental biology, but because this is an unwieldy phrase it is almost universally referred to by its nickname.

Fundamental to any field of science is the search for general, rather than piecemeal, explanations. However, we can only generalize as far as is consistent with the facts at hand. Biology is less fertile scientific ground for generalizations than physics, because there are nearly always exceptions to any proposed general rule (for example, there are exceptions to Mendel's 'laws' of inheritance). The solution to this problem for biologists is not – of course – to abandon the quest for general explanations, but rather to recognize how far any proposed generalization can go, and where its limits are set.

Against this background, we should consider the scope of evo-devo, and of any proposed general explanations that emerge from this relatively new scientific discipline. I said in the opening paragraph that evolution and development are 'intrinsically interwoven'. However, while this is true for parts of the living world, it is not true of others. If we restrict 'development' to multicellular organisms in which changes occur in populations of self-adhering cells, such as embryos, larvae, or juveniles, then life forms that are unicellular throughout their entire life cycle do not have development as such. For example, a bacterium that lives for a certain period as a single metabolizing cell and then splits into two identical daughter cells by asexual reproduction cannot be said to have a developmental phase in its life cycle. In contrast, a human, a dolphin and a butterfly most certainly do have a developmental phase – indeed the butterfly has three of them (embryogenesis, larval growth, and metamorphosis).

Although I have contrasted bacteria with animals to make this point, the difference between the realms of life that are characterized by (a) occurrence of development and (b) absence of development is more complex than this introduction to the subject suggests. The realm to which development (and hence also the evolution of development) applies is that of multicellular organisms – or, to put it more precisely, the realm of organisms that go through at least one multicellular phase in their life cycles. This means that evo-devo concerns itself not just with animals but also with plants, and with some members of other kingdoms – for example the fungal and brown-algal kingdoms. It also potentially deals with certain 'microbes' (an undefined but generally useful term) – the ones that have a phase in their life cycle that is multicellular, albeit relatively simple, such as those cyanobacteria (previously called blue-green algae) that can form filaments or mats of attached cells.

At this point it becomes helpful to have some idea of the broad-scale structure of the living world, in terms of its hierarchical division into its three major groups (called domains) and the major subgroups within these (called kingdoms). Our view of broad-scale structure has changed considerably over the last few centuries. It will continue to change in the future, but probably much less than in the past, assuming that we are gradually homing in on a correct understanding of the course that evolution has taken. Figure 1.1 shows the broad structure of the living world, as currently perceived. Development and evolution of development characterize one of the eukaryote kingdoms in its entirety (animals), most of another (plants, defined to include both green algae and land plants), and parts of at least two others (fungi, which includes unicellular yeasts as well as multicellular toadstools, and what I think of as

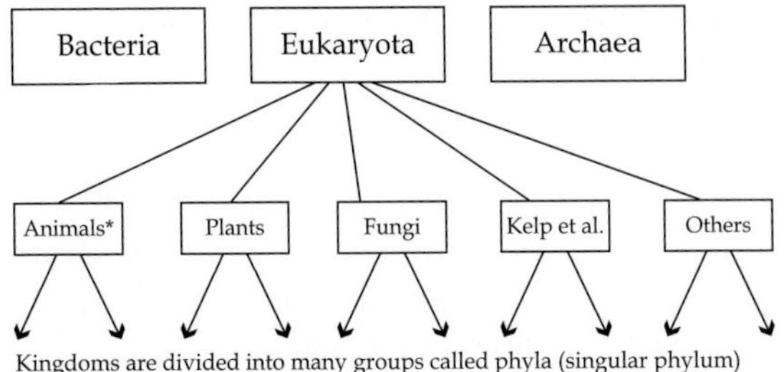

Figure 1.1 The broad structure of the living world: three domains, each divided into kingdoms. Here, kingdoms are only shown for the eukaryote domain. Evo-devo applies to animals, to most plants, and to some members of other kingdoms.

'the kelp kingdom', which includes unicellular diatoms as well as multicellular brown algae). Other eukaryote kingdoms, and the domains of Bacteria and Archaea, are not entirely without development, but its occurrence is very patchy, and evo-devo to date has largely omitted consideration of members of these groups.

Evo-devo began in the early 1980s with studies on animals, and consequently the evo-devo of animals is better known than that of other relevant kingdoms, with the evo-devo of plants coming second. Partly because of this, and partly because I am a zoologist and know the animal kingdom better than I know the others, animal examples will predominate in this book. I hope the reader will forgive this bias, and in mitigation I can at least say that the case studies discussed will range widely across both vertebrates and invertebrates.

Although evo-devo is inapplicable to some life forms on Earth, it may well in the future turn out to be applicable to many life forms elsewhere. At the outset of his 1992 book entitled *Natural Selection*, the American biologist George C. Williams states a philosophical position: he believes that natural selection will be seen to characterize all life forms in however many biospheres exist in the universe – probably a very large number. Similarly, I will state a philosophical position here: evo-devo will be seen to be relevant to all life forms everywhere that are multicellular in their construction. This is a slightly different sort of statement, since natural selection is a process while evo-devo is a field of study. However, I would predict that the most important processes involved in the evolution of development here, as discussed in later chapters, will be relevant on other planets too.

We've now considered the realm within which evo-devo studies are meaningful. Or, in other words, we've clarified the realm in which evolution and development are interwoven. Having done that, we now turn to the *way* in which they are interwoven. From here on I will focus on the animal kingdom, unless specified otherwise. However, many of the points made will be equally applicable to plants, and to other multicellular organisms.

The development of any animal can be thought of as a trajectory from zygote to adult. Since I will be using 'developmental trajectory' often in this book, I should explain here what I'm thinking about when I use this term. Imagine the development of a human. Each of us starts our life as a single cell, which starts to multiply, producing a self-adhering cluster or ball of cells. As it continues to grow, this cluster begins to take a more definite shape, with elongation in one direction producing the anteroposterior (or head-to-tail) axis of the body. Internal changes, such as the separating out of different tissue layers, accompany the external changes in shape, and the overall growth. The embryo gradually elaborates its features, becoming more and more like a miniature human. After birth, development continues, but is more subtle. One important aspect of the post-embryonic development of a human is differential growth rates of different parts of the body, something that is referred to as allometric growth (to distinguish it from isometric growth, where different body parts grow at the same rate). For example, our heads grow more slowly, after birth, than our trunks and limbs. The combination of all these changes leading from zygote to adult constitutes the developmental trajectory of a human.

At any moment in evolutionary time – say halfway through the Jurassic period – the developmental trajectory found in a certain kind of animal – say a particular species of dinosaur – has been produced by the accumulated evolution of the past and is the starting point for the evolution of the lineage concerned in the future. Development is a quasi-predictable process. Its many repeat occurrences within a given species produce variants that are typically rather similar to each other – though not identical. In contrast to development, evolution is a very *un*predictable process. The fact that one dinosaur developmental trajectory gave rise to all of today's 10,000 species of birds while the others left no descendants among today's fauna could not have been foretold. Evolution incorporates a major element of 'historical contingency' – chance events including asteroid impacts and volcanic eruptions – as emphasized by the American palaeontologist Stephen Jay Gould. Development is usually much less affected by such contingency.

One way of looking at the intertwining of evolution and development, then, is that development is a sort of raw material that gets moulded by natural selection in ways that adapt it to the prevailing environmental conditions in the relevant habitat. For example, most frogs, including all those living in temperate habitats, have a process of indirect development that includes a tadpole stage, but some species living in warm, moist, tropical conditions have dispensed with the tadpole stage of this ancestral life cycle and have evolved a process of 'direct development' (the embryo gives rise directly to a juvenile that's like a small adult). Among the many species of frogs living in temperate regions, evolution has again moulded the pattern of development to fit the environment, but in less dramatic ways. For example, a shorter tadpole phase of the life cycle would be expected to be favoured by selection in regions where the water bodies inhabited by the tadpoles are more transient than they are in others.

But the interweaving of evolution and development is not a one-way street. Evolution alters the developmental process, to be sure. But the evolutionary process is also altered by development. Or, to put it another way, evolution is effectively channelled in terms of what it can do with a particular lineage in the future by the prevailing developmental trajectory of the species concerned in the present. Likewise, evolution at any point in the past – say the mid-Jurassic again – was channelled by the developmental trajectories that were available at that time. This channelling is often referred to as 'developmental constraint'. However, I think this phrase has an overly negative tone. If development in some sense channels the direction of evolution, then it can be thought to steer it towards some types of change just as much as it steers it away from others. We'll look at various examples of this steering effect, which can be called developmental bias, in subsequent chapters.

The practitioners of evo-devo are not a homogeneous bunch. Those who are above a certain age have migrated there from various disciplinary backgrounds, because when they were students the field did not exist – or perhaps had just begun but was not yet widely studied or well funded (regrettably the last of these remains true, though the situation is a little better than it used to be). Some practitioners have come from developmental biology, some from evolutionary biology, and some from elsewhere. Partly because of their heterogeneous origins, these practitioners are also rather heterogeneous in their views of the nature of evo-devo. With regard to the two sides of the interaction between evolution and development noted above, some emphasize one, some the other, and some both. This heterogeneity of views is of interest in relation to the wider philosophy of evo-devo. However, before considering the philosophical aspects of the new discipline, we need to know a little more about it – and that includes understanding its origins.

Origins of Evo-Devo: the Homeobox

If you were to ask me, 'what was the single most important discovery in the origin of evo-devo?' I'd reply, with little hesitation, 'the homeobox.' So that's a good place to start. This was a discovery made at the same time (in 1983, with publications following in 1984) by researchers in two laboratories – that of Thomas Kaufman in Bloomington, Indiana, and that of Walter Gehring in Basel, Switzerland. The lead authors of the papers concerned were Matthew Scott and William McGinnis.

The key question at this point is: what is a homeobox? To answer this we need to start with the related question: what is a gene? A reasonable working definition is that a gene is a stretch of DNA that's thousands of nitrogenous bases long, and that makes a particular product (typically a protein). Each different gene makes a different protein, because each gene is a unique sequence of the four bases that we're familiar with by their initial letters of A, C, G, and T (full names adenine, cytosine, guanine, and thymine). Recall that the genetic code works in triplets, so three bases in a DNA strand give rise to one amino acid in the resultant protein. For example, the sequence AAA in a gene corresponds to the amino acid lysine in the protein. If we say that a typical protein is 333 amino acids long (just a rough guess), then the gene making it must be at least 999 bases long. In fact, it is normally much longer than this for various reasons, principally that the genes of organisms from the kingdoms that have development (animals, plants, etc.) typically contain stretches of DNA (called introns) whose RNA counterparts are chopped out before the protein is made. So the typical animal or plant gene is in fact thousands of bases long, rather than hundreds.

Now we return to the homeobox. A 'box', in genetics, is simply a rectangle drawn around a certain stretch of DNA to highlight it, for whatever reason. For example, if I wanted to highlight the AAA stretch in the following sequence to show you which bit of a longer sequence coded for the amino acid lysine, I'd draw a box around the central three bases (here I'm using bold text to do the same thing): TAT**AAA**GGG. The homeobox is a much longer sequence of bases than this. It is typically 180 bases in length, thus corresponding to a sequence of 60 amino acids in the corresponding protein – which in turn is called the homeodomain. In a homeobox-containing gene whose overall coding sequence is 1800 bases long, the homeobox is 10% of the gene (Figure 1.2), whereas in a gene that is 18,000 bases long (perhaps due to multiple long introns), the homeobox is just 1% of the gene's complete DNA sequence. In a homeodomain-containing protein that is 300 amino acids long, the homeodomain itself makes up 20% of this overall length.

So far, we recognize a homeobox as a sequence of a particular length that can be found within certain genes. But what *is* the sequence, which genes is it found in, and why is its inclusion in these genes significant?

It's not possible to specify the 180-base homeobox sequence *exactly*, because there are many variant versions of it. It's a recognizable pattern, or 'motif', rather than a precise sequence. Typically, one variant will be the same as another for most of the 180 bases, but will differ in a minority of them. Part of this variation is due to the redundancy of the genetic code. For example, it's not just AAA that codes for the amino acid lysine, AAG does so too. Thus it's possible to get a homeodomain with lysine in a particular position, by having either of these triplets in the corresponding stretch of the homeobox that codes for it.

But there is variation in the exact amino-acid sequence of homeodomains too. One variant homeodomain will typically be the same as another for most but not all of its amino acid sequence. The most important thing is that regardless of some variation in the structure of the homeodomain at this level, at a higher level its overall 3D structure is maintained. This 3D structure is much more complex than shown in Figure 1.2, which is schematic. We don't need to know this structure in detail, but its key feature is that it has three helical

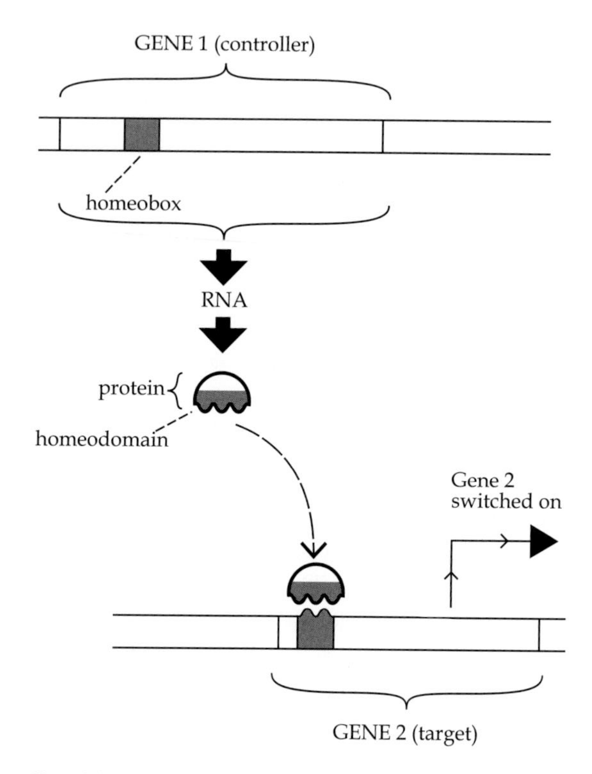

Figure 1.2 Diagram of a homeobox gene and the homeodomain protein that it makes. The homeobox represents only a small fraction of the overall length of the gene. Likewise, the homeodomain represents only a fraction of the protein. The important point to note here is that the protein's homeodomain enables it to bind to the DNA of other genes and to switch them on, thus causing them to make their own protein product. This provides a basis for developmental patterns of gene activity in the embryo.

regions that are vital to its DNA-binding function, and that are conserved in all variants.

Now let's turn to the genes in which a homeobox sequence is found. Genes can be roughly divided, in terms of their function, into three main categories: developmental (crucially important here), cell-type specific (less important), and housekeeping (least important in evo-devo). As the names of these rough-and-ready categories imply, the first make proteins that contribute to the developmental process, the second make proteins that are restricted to certain cell types (such as the haemoglobin of our red blood cells), and the third make other proteins (often enzymes) that contribute to the housekeeping tasks that occur within almost all cells, such as the metabolic activities that keep a cell supplied with energy. As usual in biology, categories are never clear-cut, but this way of thinking about gene function is helpful nevertheless – and we'll delve further into it in Chapter 4.

Many developmental genes contain a homeobox; in contrast, other genes generally lack this sequence. There is a good reason for this difference. A major part of the causality of development is cascades of gene activity in which the product of gene A switches on gene B, whose product switches on gene C, and so on. To switch on a gene, a protein must bind to the DNA of that gene. And it turns out that the homeodomain is a DNA-binding region. So if we discover a new gene about which we initially know nothing, sequence it, and discover that it contains a homeobox, we are pretty sure that it plays a role in the development of the animal concerned.

The animal in which the homeobox was discovered was the fruit-fly *Drosophila melanogaster*. The genes in which it was discovered were already known from the fact that mutations of them produce bizarre adult flies that have legs growing out of their heads or two pairs of wings instead of the normal single pair. Back in the late nineteenth century, these mutations were called homeotic, and the phenomenon they produce was called homeosis – the right thing in the wrong place. This contrasts with the wrong thing in the right place, which is a much commoner type of mutation. An example of the latter in *Drosophila* is the vestigial-winged mutant, where the wings are in the right place but are small and malformed. The person who coined this terminology (homeosis/homeotic) was the British geneticist William Bateson, whose book *Materials for the Study of Variation* was published in 1894. We'll come back to him in Chapter 2, when we look at the antecedents of evo-devo. For now, it's just important to know that it was from his homeosis that the homeobox sequence got its name.

The huge significance of the homeobox arises from the fact that it was found to characterize multiple developmental genes and, as research continued in the 1980s, it was found to characterize such genes not just in flies but in animals generally. This suggested that the genes that contributed to the developmental process were similar even when the end result of the process – the adult animal – was very different. In other words, it suggested that we could generalize about some aspects of the causality of development right across the animal kingdom, and perhaps even beyond. To say that this was exciting stuff would be an understatement. However, one discovery does not by itself make a new scientific discipline. So let's now look at some important events that preceded and followed the discovery of the homeobox.

Origins of Evo-Devo: Other Factors

In 1977, Stephen Jay Gould rekindled interest in the relationship between evolution and development with the publication of his book *Ontogeny and Phylogeny* (which, with a few ifs and buts, simply means Development and Evolution). Two years later, he and fellow Harvard biologist Richard Lewontin wrote an influential – and controversial – paper championing the role of various forms of constraint in evolution, including developmental constraint, and downplaying the role of natural selection. We'll discuss this paper, and the concept of constraint, in Chapter 4.

In 1978, the American geneticist Ed Lewis published an important paper on the structure of a gene complex in *Drosophila* that contained multiple genes that were subject to homeotic mutation and that would, a few years later, be found to contain homeoboxes. In 1980, the Heidelberg-based biologists Christiane Nüsslein-Volhard and Eric Wieschaus published an equally important paper on other genes that contribute to the development of *Drosophila*, including the now-famous *hedgehog* gene that gives mutant larvae a prickly appearance – hence the name. The genes studied by these three biologists were similar in that they all affected the development of segments – those sections of the body along the head-to-tail axis of which an insect is made. However, in another way they were different. The genes studied by Lewis were involved in the determination of segment identity (e.g. thoracic vs. abdominal). In contrast, those studied by Nüsslein-Volhard and Wieschaus were involved in the determination of segment number and segment polarity (which end of a segment is anterior and which posterior); these authors showed that such determination took place in stages of increasingly restricted spatial scope, starting with the whole body and culminating in the patterning of individual segments. All three of these biologists shared a Nobel Prize in 1995 for their work on developmental genes, which helped to pave the way for later comparative studies.

Through the rest of the 1980s, the 1990s, and into the present century, work on comparative developmental genetics expanded rapidly. Other developmental genes than those subject to homeotic mutations were discovered to contain homeoboxes, so the relationship between homeobox genes and homeotic genes became more complex than the one-to-one relationship that might at first have been thought. And many developmental genes were discovered *not* to have homeoboxes, which suggested that some of them might have a different developmental role than making proteins that directly switch other genes on. For example, the *hedgehog* gene makes a protein (called Hedgehog) that is secreted out of a cell, transported in the intercellular matrix, and bound by a receptor protein on the surface of another cell. Hence it is a cell–cell signalling protein – one of a number of such proteins that make up what is called a signalling pathway. This is important, because development involves not just the turning on of certain genes in certain cells but also the coordination of such switching events between cells in space and time, both in embryos and in post-embryonic developmental stages.

Although certain genes are crucially important in the developmental process, I have deliberately avoided using the statement 'genes control development', which regrettably can be found in many publications. Such statements are rightly criticized as smacking of what is sometimes called genetic imperialism. In fact, development is a two-way interaction between genes and other molecular players, including proteins. Genes make proteins, while certain proteins control the activities of genes; and this interplay is often influenced by environmental factors. In a collaborative study by members of Michael Akam's laboratory in Cambridge and my laboratory in Galway, it was shown that in contrast to the situation in flies, where the number of segments is not influenced by the environment, the number of segments in some species of centipedes can be altered by the temperature at which embryos develop – this finding was published by Vincent Vedel and colleagues in 2008. A year later, the American biologists Scott Gilbert and David Epel published their book *Ecological Developmental Biology* (second edition 2015), which emphasized the importance of the environment – including other species – in influencing the route that development takes. The nickname eco-devo is now sometimes used for such an approach.

Evo-Devo and Darwinism

In *The Origin of Species*, first published in 1859, Darwin famously described natural selection as 'the main but not exclusive means of modification' in the evolutionary process. While some version of this statement is probably true, it's not as simple a statement as it first seems. To see why, we need to discuss levels of biological organization. These extend from the molecular level up to that of the cell, then to tissues and organs, then to the individual animal or plant, and finally (for our purposes here) up to the population – the collection of all those individuals of a particular species living in a particular area.

Natural selection is a population-level process. At this level, selection probably is, as Darwin suggested, the main means of evolutionary change. Perhaps the only evolutionary theory of the last half-century or so that denies this is the neutral theory, proposed by the Japanese geneticist Motoo Kimura in the 1960s and explained at length in his 1983 book *The Neutral Theory of Molecular Evolution*. Kimura argues that, with certain provisos, more of the evolutionary changes in genes and their protein products are driven by 'genetic drift' than by natural selection. Drift is the name given to the random fluctuations in a population's genetic make-up that are most noticeable when selection is absent. In the short term, when applied to a single gene, these don't achieve much; however, in the long term, across many genes, drift will sometimes cause many changes in the same direction, just by accident. It's like asking all the students in a lecture room to toss a coin ten times. If there are enough students, someone will get ten heads in a row just by chance.

However, the 'certain provisos' mentioned above are important. While Kimura questioned Darwin's theory at the level of individual genes, he acknowledged that, in terms of organismal body form, natural selection prevails. So, the long neck of a giraffe is not a result of genetic drift any more than it is a result of the giraffes' neck-stretching being passed on to their offspring (the so-called 'inheritance of acquired characters', for which there is no evidence).

Evolution is a multi-level process, not just a population process. It involves molecules, cells, and organisms, as well as populations. What happens when we simultaneously consider more than one level? Here are two alternative multi-level scenarios.

1. Suppose that many mutationally induced changes to development are possible, and that they are all equiprobable. If a new variant is fitter in a general sense, as summed up over all stages of the life cycle, it may spread under the influence of natural selection in the population. In such a scenario, there is no conflict between developmental and population levels of organization.
2. Suppose instead that some variant patterns of development are inherently more probable than others in the sense of being more easily produced (this can be called developmental bias, arguably a better term for developmental constraint, as noted earlier). Bias could arise because of some aspects of the dynamics of the developmental process. If it is strong enough, bias in favour of one variant may be more important than natural selection in favour of the alternative variant.

The Canadian biologist Brian Goodwin suggested the second scenario in the case of leaf-development patterns in flowering plants (the angiosperms: a group of more than a quarter of a million species). There are three main such patterns: (a) leaves branching off a stem on alternate sides (distichous); (b) leaves branching off in pairs or groups, with each group being rotated through 90 degrees with respect to the previous one (decussate); and (c) each leaf branching off at a certain angle to the one before it (spiral). In his 1992 book *How the Leopard Changed its Spots*, Goodwin notes that more than 80% of angiosperm species have the spiral pattern of leaf development. And he goes on to suggest that this is because the spiral pattern is the most easily generated by plant developmental systems. He says (Chapter 5, page 132): 'the frequency of the different phyllotactic patterns in nature may simply reflect the relative probabilities of the morphogenetic trajectories of the various forms and have little to do with natural selection.'

There are, however, at least two problems with Goodwin's suggestion. First, the significance of 80% of angiosperms having a spiral pattern of leaf development depends on how they are scattered across the angiosperm evolutionary tree. The spiral developers could have had just one origin, and characterize a single but very large branch. Alternatively, they could have arisen lots of times and characterize instead multiple small twigs. Goodwin did not deal with this aspect of the problem, and perhaps in retrospect he was wise – because the perceived structure of the angiosperm phylogenetic tree changed dramatically with the blossoming of plant molecular phylogenetics shortly after his book was published.

The second problem is this: it's hard to imagine that the pattern of arrangement of the leaves of an angiosperm would not be subject to natural selection. After all, leaves are the major energy-acquisition organs of flowering plants (with some exceptions, such as cacti). The pattern in which they develop and hence are arranged thereafter must be of major importance in terms of minimizing leaf overlap and maximizing photosynthetic activity. However, Goodwin suggests (on page 132 again) that 'all the phyllotactic patterns may serve well enough for light-gathering by leaves and so are selectively neutral'. It is one thing to suggest, as Kimura did, that a single amino-acid change in a protein that contains hundreds of amino acids may be selectively neutral; it is quite another thing to suggest that the overall structure of a plant's main photosynthetic apparatus is neutral.

Clearly, the philosophical stances associated with advocating the two scenarios described above are very different. Number 1 is essentially 'neo-Darwinian' while number 2 could be described as 'anti-neo-Darwinian'. But there is an intermediate scenario, with an associated intermediate philosophical stance, which will gradually come into focus as we proceed. *In this scenario, developmental bias and natural selection interact with each other to produce evolutionary trends in particular directions.* Many practitioners of evo-devo are attracted by this idea. (An interesting source of information on the role of developmental bias in evo-devo thinking is provided by a 2020 special issue of the journal *Evolution & Development*, edited by Armin Moczek.)

However, many practitioners of evo-devo are not overtly philosophical in their approach. I would describe the molecular evo-devo that began in the 1980s as a largely practical endeavour, in which the focus was on doing more case studies, accumulating more information in general, and waiting to see what generalizations suggest themselves when we have a more extensive database. Today, much evo-devo is still of this data-gathering flavour. It is particularly important in this endeavour that our data are well spread across the animal and plant evolutionary trees. This can be referred to as good taxon sampling, both in the sense that no major branches are omitted and in the sense that we can make multiple comparisons at all levels of taxonomic distance, from closely related species (e.g. those belonging to the same genus or family) to representatives of different phyla.

Despite its current practical focus, evo-devo is not a philosophy-free zone. Indeed it has attracted the attention of many philosophers of science, who are interested in the impact that evo-devo may have on our understanding of evolution in general. Of course, it might turn out that many evo-devo studies are just filling in some missing details. Perhaps this is true of those studies that revealed some of the developmental genes and proteins that underlie the differences in beak size and shape among different species of Darwin's finches, without altering the basic adaptive scenario under which the beaks are thought to have evolved. Then again, it may turn out that some evo-devo studies are really saying something new and important about how evolution works. This is perhaps most likely to be the case in relation to the origins of evolutionary novelties and body plans (e.g. the turtle shell and the vertebrate skeleton respectively). We will deal with these origins from various perspectives in Chapters 6, 7, and 8.

The Importance of Evo-Devo

Evo-devo has a general importance that transcends the issue of whether it does or does not end up producing a new view of the ways in which evolutionary novelties and body plans originate. Our new discipline is essentially putting the individual animal or plant back at centre-stage in our overall perspective on the pattern and process of evolution.

In the couple of decades before evo-devo started, much of evolutionary theory was focused on two particular levels of organization – molecules and populations. Strangely, the intermediate level of the individual organism was neglected. The prevailing approach in the field of population genetics was that we could be dealing with a barnacle or an elephant – it didn't really matter. Although this approach now seems bizarre, it arose for a good reason. It's important to understand this reason, because it helps us to put the evo-devo endeavour that came later (the 1980s) into a broader evolutionary context. And to understand why molecular population genetics became so dominant in the 1960s and 1970s, we need to look back a couple of decades earlier than that.

In the 1940s and 1950s, evolutionary case studies carried out in the wild were typically based on phenotypic variation in natural populations that was visible to the naked eye. This was because many of the molecular methods that are now used to 'see' things – like genes and proteins – inside organisms did not yet exist. Some studies focused on continuous variation – such as that in the size of a snail's shell. However, the interpretation of such studies was complicated by the fact that continuous variation is only partially inheritable. Other studies focused on polymorphic variation, that is, the type of variation in which there is a small number (sometimes just two) of phenotypes or 'morphs' that are clearly distinct from each other and do not intergrade via a long series of intermediates. An example is the case study of different colours and banding patterns on the shells of various species of land-snails, notably *Cepaea nemoralis*, and the way in which the relative frequency of these morphs varies from place to place. This latter type of study had the advantage that breeding experiments revealed the discrete phenotypes to be completely inheritable and controlled by particular genes. However, they had the disadvantage that most animals do not exhibit polymorphic variation that is externally visible. Hence the case studies were restricted to groups that do – land-snails, lepidopterans, ladybirds, and a few others. Their broader relevance was uncertain.

In the 1960s everything changed. The introduction of a particular technique (gel electrophoresis) allowed the screening of natural populations of any type of organism for genetic variation in the many genes that produce enzymes. Pioneering work on populations of flies and humans revealed that such genes were often polymorphic. Since all organisms have multiple enzyme-producing genes, this work could be (and was) extended to many taxa. Unlike the earlier work on pigmentation polymorphisms, the work on enzyme polymorphisms was clearly applicable across the living world. And the main discovery was that polymorphism was widespread. By the 1970s, this had been established beyond reasonable doubt.

This discovery was followed by a heated debate over whether variation within and between populations was acted upon by natural selection, or was selectively neutral and hence only acted on by the random process of genetic drift. The answer turns out to be 'a bit of both', with the relative importance of the two still not entirely clear. Two of the key books about this debate are Richard Lewontin's *The Genetic Basis of Evolutionary Change*, published in 1974, and Motoo Kimura's 1983 book *The Neutral Theory of Molecular Evolution*, which I mentioned earlier.

With the benefit of hindsight, this work of the 1960s and 1970s can be thought of collectively as being 'the evolutionary genetics of housekeeping genes'. Many of the enzymes studied were standard players in metabolism, for example enzymes involved in the energy-generating processes that are common to most cell types, such as glycolysis (liberation of energy from glucose). Hence the loss of focus on the organism and its body form. The developmental trajectories of a barnacle and an elephant are very different, but their main metabolic pathways are much the same.

In the 1980s, interest shifted in two directions. Studies of molecular evolution shifted from gene products – proteins – to the genes themselves, and these studies branched out to incorporate evolutionary change in DNA sequences that are not 'standard genes' – e.g. repetitive DNA, transposable elements, and pseudogenes. (To learn more about that work, you will need to read an introductory book on molecular evolution, as it's outside the scope of this one.) At the same time, evo-devo arose from discoveries such as the homeobox, which allowed us to examine not housekeeping genes but developmental genes – the genes that build the organism.

So, the importance of evo-devo, at its most general, is that it has put the organism back at the centre of evolutionary thinking. We are not just interested in the evolution of an animal's metabolism, we are also interested in the evolution of its *form*. And indeed perhaps we should be *more* interested in the latter, though of course this is a subjective opinion. It is an opinion that was nicely put by the British biologist Conrad Hal Waddington (whom we'll meet again in Chapter 2). In 1975 – shortly before the advent of evo-devo – Waddington perspicaciously said this:

> It is doubtful if anyone would ever have felt any need to resist the notion of evolution if all it implied was that the exact chemical composition of haemoglobin changed over the ages.

Groups of people resisting the idea of evolution – biblical fundamentalists and other creationists – do so on several bases. One is the unacceptability (to them) of evolutionary theory's refutation of the creation myth featuring Adam and Eve, some version of which appears in all the Abrahamic religions – Judaism, Christianity, Islam, and several other smaller ones. Another is the notion of the supposedly 'irreducible complexity' of biological systems – the idea that these systems are too complex to have arisen by the blind processes of mutation and selection. While molecular complexity (e.g. Waddington's haemoglobin) is in fact objected to as a product of evolution by creationists, what they object to much more is the idea that objects as impressively complex as the teeth of a tyrannosaur or (especially) the brain of a human are also products of evolution.

We can link this argument about what creationists object to the most with the question of what would excite biologists the most in terms of the discovery of life on other planets. Finding a biosphere consisting entirely of microbes would be very exciting – and such a discovery may only be a few decades away (via studies of exoplanet atmospheres). But finding a biosphere with large multicellular creatures characterized by complex developmental systems that produce, among other things, sophisticated brains, would be more exciting still. It is evolution's pinnacles of achievement that excite us most of all.

But let me finish with a caveat. Evolution is not an escalator. There is no such thing as a universal upward trend in brain size (or anything else), as we will discuss further in the next chapter, in terms of the old idea of a *scala naturae*. In the lineage from an ancient worm to a modern human, brain size started small and then got bigger and bigger, but this trend was interrupted by frequent long periods of stasis and perhaps also by shorter periods of decline. One of the most closely related animal phyla to the vertebrates is the Echinodermata – the group that contains the starfish, sea-urchins, and sea-cucumbers, among others. In the lineage from that same ancient worm (the last common ancestor of vertebrates and echinoderms) to a present-day starfish, brain size went from small (possibly via a bit bigger) to zero. We always need to keep the messiness of evolution at the back of our minds. Evolution isn't 'trying' to achieve anything. It simply happens. It's an inevitability of having the three linked properties of variation, reproduction, and inheritance. And the evolution of development is an inevitability of having those three plus the fourth property of having a multicellular body with a developmental trajectory.

2 Antecedents of Evo-Devo

Quasi-Distinct Stretches of Time

Although today we call the scientific study of the relationship between evolution and development 'evo-devo', neither that term, nor its longer counterpart 'evolutionary developmental biology', existed before about 1980. Yet the study of the relationship between the two great creative processes of the living world has a much longer history – effectively starting in the nineteenth century, the first century in which there was a well-articulated theory of evolution (first Lamarck's, then Darwin's). We generally refer to evo-devo's nineteenth-century antecedent as 'comparative embryology'. Although in the subsequent period from about 1900 to 1980 there were further studies of the relationship between evolution and development, there is no collective term for this endeavour, because mainstream developmental biology and evolutionary biology were largely separate undertakings during that stretch of time. The few biologists who tried to deal with the two together over this 80-year period might be described as mavericks. Each of them produced interesting bodies of work, but these did not really link up to form a scientific discipline.

We'll examine these two antecedents of evo-devo – the comparative embryologists and 'the mavericks' – in turn. But before we do so, we should ask the following question: Is it really true that nothing of interest to current evo-devo happened before 1800? The short answer is 'no'. There is not space in this book for a detailed discussion of the period before 1800, but a few brief comments, constrained within a single paragraph, may be helpful.

Charles Darwin's grandfather Erasmus proposed an evolutionary view of the living world in the 1790s – but without specifying an evolutionary mechanism. The Italian anatomist Hieronymus Fabricius published an important work describing his pioneering studies on embryology at the start of the seventeenth century. The French naturalist Pierre Belon wrote a book on the natural history of birds in the 1550s, in which there is a famous and much-reproduced drawing showing the correspondence of the bones in a bird skeleton with their counterparts in a human skeleton, which anticipated the evolutionary concept of homology (see next chapter). And among the ancient Greeks there may well have been a few evolutionists. Erasmus Darwin says as much, but regrettably without giving details or names. We know that Aristarchus of Samos came up with the idea of a heliocentric solar system almost two millennia before Copernicus. Given that, perhaps scholars of the classical world will one day discover evidence of an ancient Greek evolutionist to back up Erasmus Darwin's claim.

So, in our search for antecedents of evo-devo, we can think in terms of three stretches of time: pre-1800 (the above paragraph), the nineteenth century (next section), and the period from about 1900 to 1980 (the section after that). After this chapter, we return to recent and current evo-devo (1980 onwards).

Nineteenth-Century Comparative Embryology

Two of the most important comparative embryologists of the nineteenth century were Karl von Baer and Ernst Haeckel. Both conducted extensive studies on the development of many kinds of animals, and both attempted to make broad generalizations based on their studies. Unfortunately, both went further than generalizations and proposed 'laws' (four in the case of von Baer and one in the case of Haeckel) – which, as we have already noted, is a dangerous thing to do in biology. So now we should look at these laws and the extent to which they are currently thought to be true. From the perspective of doing evo-devo in the present, we need to know whether the work of von Baer and Haeckel is a help or a hindrance – or possibly both.

To do this, it is useful to start in the 1820s with the 'Meckel–Serres law', proposed by the German biologist Johann Meckel and his French counterpart Étienne Serres. Recall that these were pre-Darwinian times: *The Origin of*

Species did not appear until 1859. They were not, however, pre-evolutionary times. The French biologist Jean-Baptiste Lamarck had published his *Philosophie Zoologique* in the first decade of the nineteenth century; some biologists of that time accepted evolution, while others did not. The Meckel–Serres law needs to be seen against this background. Meckel was an evolutionist. Serres might be described as a quasi-evolutionist; he excluded humans from the evolutionary process (Alfred Russel Wallace later excluded the human mind, but not the human body).

The Meckel–Serres law stated that there was a parallel between the stages of development within any one kind of animal and the stages of evolution through which it had gone in the past. In its development, an animal went through all the stages of its evolutionary history, including, supposedly, the adult stages of its ancestors. The later embryonic stages of a particular animal were seen as being in some sense 'higher' than its earlier ones, just as some kinds of animal (e.g. mammals) were seen as 'higher' than others (e.g. fish). This is the *scala naturae* view of the living world, or the 'great chain of being', which can be traced back, via Charles Bonnet in the 1700s, to Aristotle and other ancient Greeks. Aristotle saw animals as being arranged on a vertical axis that had humans at the top, worms and their kin at the bottom. Above animals were (putative) angels, below them were plants. This view has been completely untenable for a long time, especially since Darwin and others replaced it with a more accurate one – a tree.

In 1828, von Baer published his major work, in which he outlined his four laws. Together they constituted a rejection of the Meckel–Serres law. Von Baer's four laws are given below, taken from the English translation that appears in Stephen Jay Gould's 1977 book *Ontogeny and Phylogeny*, which I mentioned in Chapter 1:

1. The general features of a large group of animals appear earlier in the embryo than the special features.
2. Less general characters are developed from the most general, and so forth, until finally the most specialized appear.
3. Each embryo of a given species, instead of passing through the stages of other animals, departs more and more from them.
4. Fundamentally therefore, the embryo of a higher animal is never like a lower animal, but only like its embryo.

In contrast to the Meckel–Serres law, which is now universally rejected, von Baer's laws are still regarded as having some truth in them, and as being a possible basis for some evo-devo generalizations – notably the idea of a developmental hourglass, which we'll come to in Chapter 4 – though of course it is now accepted that we should not use the term 'law'. However, there is also some *untruth* in von Baer's laws. Notably, we no longer recognize the concept of higher and lower animals as being valid.

Almost four decades after von Baer's magnum opus, and a few years after Darwin's, Ernst Haeckel published his own (in 1866), entitled *Generelle Morphologie der Organismen*. This was both long and dense; rather few copies were sold. However, Haeckel followed this up with two somewhat more readable books, both of which were translated into English – as *The History of Creation* and *The Evolution of Man* (published in 1876 and 1896 respectively; the former is the more readable of the two). Haeckel himself criticizes *Generelle Morphologie* in the preface to *The History of Creation*:

> The 'Generelle Morphologie' found but few readers, for which the voluminous and unpopular style of treatment, and its too extensive Greek terminology, may be chiefly to blame.

Rather than reading any of Haeckel's original works or their translations, many biologists use Gould's 1977 interpretation of them in *Ontogeny and Phylogeny*. While Gould's book is a useful guide to the work of von Baer (note that I used it as the source of his four laws above), it is not so useful as a guide to the work of Haeckel. Unfortunately, Gould paints Haeckel as a retrogressive figure – someone who represents a return to the Meckel–Serres law after its effective refutation by von Baer. In particular, Gould criticizes Haeckel for saying that the embryos of 'higher' animals go through the *adult* stages of 'lower' (ancestral) ones.

After reading Gould, but before reading Haeckel, my main thought was that embryos of descendant animals going through the adult stages of ancestral animals makes no sense, especially in the post-Darwinian era, and Haeckel was too good an embryologist to have proposed such a thing. Let's take the example of human embryogenesis and its link with our ancestral fish. Like all other land vertebrates, our ancestry was channelled through an ancient lobe-finned fish that invaded the land some 350–400 million years ago. Compared to statements of our even earlier ancestry, for example that all vertebrates evolved from some sort of primitive worm, the statement that humans have a lobe-finned fish as an ancestor can be regarded as a statement of fact rather than a hypothesis. The reason for this is that fish have fossilizable endoskeletons while worms do not. The story of our fishy ancestors and their invasion of the land has been very thoroughly researched; it is told in the 2002 book *Gaining Ground* by British palaeontologist Jenny Clack. Did Haeckel think that human embryos went through a stage that resembled an adult lobe-finned fish, or any other adult fish for that matter (say a salmon), or even a sort of generalized adult fish (which would necessarily have fins)? I doubt it.

Now we should look at what Haeckel actually said. In his book *The History of Creation*, Chapter 13 (in Volume I) is of special interest. At one point in this chapter, Haeckel acknowledges that von Baer criticized the Meckel–Serres law. However, he then states that when von Baer published his four laws (1828) there could be no 'actual understanding' of the issue, and that such an understanding needed to wait until after Darwin's publication of *The Origin of Species*. He then wanders off at a tangent and does not return to von Baer, thus leaving the issue open. Also, Haeckel appears to flip back and forth between a tree-like pattern of the living world (von Baer's embryonic divergence as interpreted by Darwin) and a *scala naturae* one (as in the Meckel–Serres law). Perhaps he was having an internal struggle with himself on this issue.

Haeckel's law is usually referred to as either the biogenetic law or the law of recapitulation (he himself called it 'the fundamental biogenetic law'). It is often succinctly stated in the form: *development repeats (or recapitulates) evolution*. But this succinct version of the law does not distinguish between repeating the embryonic stages of your ancestors (the enlightened version) and repeating their adult stages (which would be retrogressive, as Gould emphasized). The latter is nonsense; the former is often the case, subject to various caveats and exceptions.

Haeckel's *History of Creation*, like Darwin's *Origin of Species*, is a very wide-ranging book. However, like Darwin, Haeckel also wrote a book focusing on the evolution of humans. Darwin's *The Descent of Man* was published in 1871. Haeckel's *The Evolution of Man* was first published as *Anthropogenie* in 1874. Here is a quote from the English translation of 1896 (Chapter 1, page 18), in which Haeckel sounds entirely in accord with von Baer (the italics are mine):

> The fact is that an examination of the human embryo in the third or fourth week of its evolution [meaning development!] shows it to be altogether different from the fully developed Man, and that it exactly corresponds to the *undeveloped embryo-form* presented by the Ape, the Dog, the Rabbit and other Mammals, at the same stage of their Ontogeny.

The only fault in this statement is the use of 'exactly'. The early embryos of most mammals are similar to, rather than exactly the same as, each other. There we have to leave Haeckel, but anyone interested in the personal background to Haeckel's scientific work should read the excellent 2008 biography *The Tragic Sense of Life*, by the American historian of science Robert Richards.

Our conclusion from this quick foray into nineteenth-century comparative embryology is that when we compare the development of two species at various stages from early embryo to adult we find some interesting patterns, including both von Baerian (embryonic divergence) and 'enlightened Haeckelian' (imperfect recapitulation that does not include ancestral adults). Comparison of *present-day* species often gives the former pattern (Figure 2.1); comparison of a descendant species with a progenitor species often gives the latter – e.g. human embryos having gill clefts similar to those that the embryos of our Devonian fishy ancestors had). I would describe the two patterns as different ways of looking at the same thing.

However, we have not quite finished our brief nineteenth-century historical tour. The story of comparative embryology and comparative anatomy of that era was not an entirely Germanic one (Haeckel was German; von Baer was ethnically a Baltic German who lived in the Russian Empire). Biologists of several other nationalities were also involved. Regrettably, there is insufficient space for most of them. However, an exception must be made for the Frenchman Étienne Geoffroy Saint-Hilaire (often referred to simply as Geoffroy), who proposed that vertebrates are, in a sense, upside-down arthropods. (Geoffroy accepted Lamarck's evolution, while some of his contemporaries, notably his nemesis, Georges Cuvier, did not.) The basis for Geoffroy's proposal was that a vertebrate develops its main nerve cord along the dorsal midline of its body, while an arthropod develops its equivalent cord along the *ventral* midline. Other structures/organs likewise occupy approximately opposite positions on the dorsoventral axis when vertebrates and arthropods are compared.

Many biologists of the time did not agree with Geoffroy's proposal; indeed, some thought of it as rather fanciful. It rapidly receded from biological

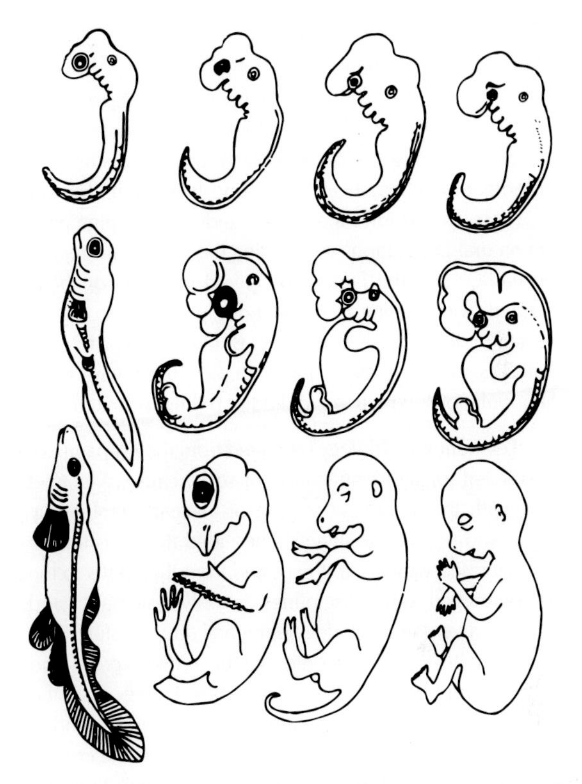

Figure 2.1 Embryos of present-day fish, hen, cow, and human, showing the von Baerian pattern of early similarity (top) giving way to later differences (bottom).

discussion. However, it made a dramatic re-entry in the evo-devo era, more than 150 years after its proposal, with the discovery that certain genes that have a dorsalizing effect in vertebrates have a ventralizing effect in arthropods and vice versa.

From Haeckel to Homeobox

For most of the period from the 1890s, in which decade Haeckel's *Evolution of Man* was published, to the 1980s, in which the homeobox was discovered, developmental and evolutionary biology went their separate ways, as noted above. In developmental biology, comparative embryology of different species was replaced by experimental embryology of individual species – in particular those that were easy to work with (this idea of workable species led to the concept of 'model systems', which we'll look at in Chapter 3). In evolutionary biology, population genetics took over as the mainstream approach, initially in terms of mathematical theory and later in the form of practical case studies. Those case studies that were conducted on pigmentation polymorphisms in natural populations – including the land-snail *Cepaea nemoralis* (mentioned earlier) and the famous peppered moth *Biston betularia* – are often described as belonging to the sub-discipline of 'ecological genetics', a term invented by the English biologist E. B. ('Henry') Ford. Many other kinds of case studies followed these, for example on enzyme polymorphisms in natural populations, which we looked at briefly in Chapter 1.

Both developmental and evolutionary biology benefited from their separation. Experimental embryology led to the identification of some causal agents of embryogenesis, starting with the 'organizer' region of the early embryonic stage called the blastula; and population genetics provided the basis for the 'modern synthesis' of evolutionary theory in the period from the 1930s to the 1950s, in which it combined with other disciplines, notably systematics and palaeontology. But in the divergence of developmental and evolutionary biology something was also lost; and it would not be rediscovered (and expanded upon) until the advent of evo-devo in the 1980s. However, in the period from the 1890s to the 1980s there were a few independently minded folk (or 'mavericks', as I called them earlier) who stood out from the crowd. It's time now to look at their work. However, we'll need to be very selective and restrict our attention to just a few key figures, as space is limited.

We'll start with the British biologist William Bateson, who is of particular significance for evo-devo because, as we saw earlier, he invented the term homeosis, which provided the basis for naming the homeobox. Bateson's *Materials for the Study of Variation* appeared in 1894. While the book's main title wouldn't win any prizes, its subtitle was very helpful in that it flagged up Bateson's central theme. Here it is: *Treated with Especial Regard to Discontinuity in the Origin of Species*. There is an important general point here. Several of the people studying evolution and development together in various ways in the period between Haeckel and the discovery of the homeobox were 'saltationists' – they believed that mutations with large effects on the development of an animal or plant were important contributors to evolution. Hence they were evolutionists but not Darwinians – at least in the sense that they didn't accept Darwin's gradualism, in other words his promotion of the dictum that *natura non facit saltum* – i.e. nature (in the form of evolution) does not proceed by a series of large leaps. Bateson's book included a compendium of instances of discontinuous variation that he thought were important to evolution, including homeotic mutations. This theme was picked up later by Richard Goldschmidt (see below).

Following the 'rediscovery' of Gregor Mendel's 1860s work on inheritance at the start of the twentieth century, there was a debate between the 'Darwinians' (also called the 'biometricians') and the 'Mendelians' about the relative importance of continuous and discontinuous variation in providing the basis for evolutionary change. The gradualist Darwinians thought that discontinuous variation was irrelevant. They took such variations to be what Darwin called 'sports': they occurred for sure, but they weren't important for evolution because they usually or even always decreased fitness rather than enhancing it. In other words, they decreased the probability of survival, reproduction, or both. In contrast, the saltationist Mendelians thought that continuous variation was irrelevant for evolution because, following Mendel's demonstration that inheritance worked in a particulate way, they considered continuous variation – for example in human height or the wingspan of a bird – to be uninherited and merely a sort of background 'noise' produced by direct effects of the environment and nutrition on the phenotype. A leading Mendelian in the early twentieth century was the Dutch botanist Hugo de Vries. He argued for a saltational origin of species based on his observations of evening-primroses (genus *Oenothera*), and in particular large-effect mutations of certain characters of these plants – for example leaf shape. Although it was de Vries who invented the term 'mutation', the variations he was studying were not caused by single-gene mutations (the modern use of the term), but rather by whole-chromosome (or whole-genome) effects.

The debate between Mendelians and Darwinians was defused by the British geneticist Ronald Fisher, who showed in 1918 that multiple genes each with smallish discontinuous effects could together produce a continuous spectrum of variation, often characterized by a normal distribution curve. Fisher went on to explain why individual large-effect mutations were (almost) always detrimental in his 1930 book *The Genetical Theory of Natural Selection* (we'll examine his explanation in Chapter 7). However, while the debate receded at this stage, and the Darwinians were essentially seen to have won it – since continuous variation was (partially) heritable after all – the saltationist approach would be re-championed by the German-American geneticist Richard Goldschmidt in the 1940s and 1950s. We'll get to him shortly, but first – in order to keep the story in rough chronological order – we need to discuss the work of the Scottish polymath D'Arcy Thompson and the English biologist Gavin de Beer.

Thompson is most remembered for his 'theory of transformations', which was articulated in his book *On Growth and Form*, published in 1917 (with a much-enlarged second edition in 1942). The essence of his idea of transformations is as follows. If you draw the outline of a type of animal, for example a crab, on a sheet of what might be called 'rubber graph paper', and then distort the grid in a systematic way, as shown in Figure 2.2, the shape seen is often a good approximation to the shape of related species – in this example other types of crab. Thompson's work is widely regarded as inspirational, but exactly what conclusion it inspires about the evolution of development is not clear. Thompson was interested in mathematical elegance, much less so in physical causes. One reading of his transformations is that development often changes in a coordinated rather than piecemeal way in evolution, but exactly how it does so Thompson did not address. Also, his work is made harder to interpret because his transformations were not set against the background of a particular evolutionary tree. Indeed, he typically compared one extant species with another (an evolutionary transformation that never happened) rather than an extant species with its ancestor.

Although Thompson's method of transformations is based on continuous variation – because shape distortions range from subtle to drastic with all shades of intermediates – he also emphasized that each series of transformations was limited in its taxonomic scope. For example, you can distort the outline of one crab in such a way that you obtain the shape of another crab, and you can do the same for fish, but you cannot transform a crab into a fish or vice versa. There are in some sense morphological realms within which gradual change by transformation is possible but between which it is not.

Reading what Thompson says about what his transformations cannot do is perhaps even more interesting than reading about what he says they can do, because it becomes clear that in terms of large-scale evolution he was a saltationist (e.g. second edition, Chapter 17, pages 1094–1095):

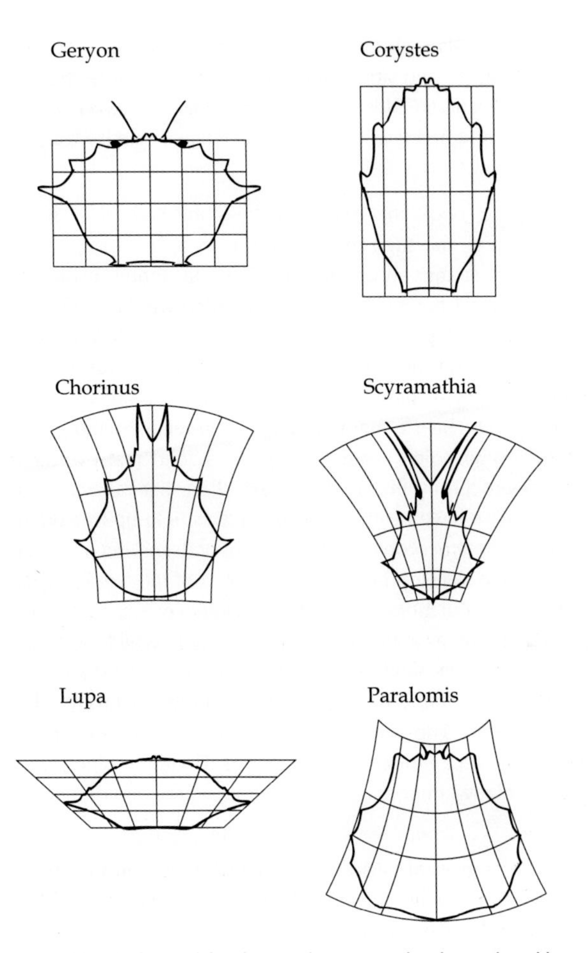

Figure 2.2 Outlines of the shapes of six types of crab, produced by applying Thompsonian transformations (see text) to a starting point of one of them (*Geryon*).

> Our geometrical analogies weigh heavily against Darwin's concept of endless small continuous variations; they help to show that discontinuous variations are a natural thing, that 'mutations' – or sudden changes, greater or less – are bound to have taken place, and new 'types' to have arisen, now and then.

However, he did not speculate about the physical causes of such 'sudden changes' any more than he did about the causes of the transformations themselves. In contrast, Richard Goldschmidt wrote extensively on the causes of sudden changes, but most biologists are now agreed that he got it wrong, as we will shortly see.

Gavin de Beer wrote a book that was initially published in 1930 as *Embryology and Evolution* and then republished in 1940 as *Embryos and Ancestors*. This was a strange book for its time. Rejecting Haeckel's recapitulation was a central theme, while little attention was paid to what were then more recent developments – for example Thompson's transformations, which only get a brief mention. Much of the book is given over to discussion of various forms of heterochrony – evolutionary alterations in developmental timing. De Beer's main negative (anti-recapitulation) and positive (pro-heterochrony) points were related. He argued that evolution involved so much heterochrony that it obliterated all vestiges of recapitulation that might otherwise have been observed. He was right up to a point: there is indeed much heterochrony, and this temporal shifting of some things relative to others in development makes the idea of the fixed 'stages' envisaged by some recapitulationists untenable. However, he came too close to seeing heterochrony as a kind of law of the way in which development evolves. This was a mistake. As the American biologist Rudolf Raff said in his 1996 book *The Shape of Life*, 'It's Not All Heterochrony' (this was his title for Chapter 8). The evolution of development involves many things, heterochrony and recapitulation among them. There are differences of opinion as to their relative importance, but over-elevating any of them is unhelpful.

Returning to the theme of gradual versus saltational evolution, in the 1930s and 1940s Richard Goldschmidt studied the homeotic mutations in insects that Bateson had drawn attention to in the closing years of the nineteenth century. He considered that these could at once change an animal so much that taxonomically it would belong not just in another species but also in another higher taxon – e.g. a family or order. For example, in addition to the homeotic mutation that we met in Chapter 1, in which antennae are transformed into legs (antennapedia), there is another such mutation in which the little flight-balancing structures called halteres that are normally found attached to the third thoracic segment are transformed into wings (bithorax).

This latter mutation could be thought of as removing the mutant flies involved from the order Diptera – which, as its name suggests, is typically characterized by possession of only two wings – and placing them in another (four-winged) order. Goldschmidt was uncompromising in his view of the general importance of such mutations in evolution, as illustrated by a statement in his 1940 book *The Material Basis of Evolution*:

> Microevolution does not lead beyond the confines of the species, and the typical products of microevolution, the geographical races, are not incipient species. There is no such thing as incipient species. Species and higher categories originate in single macroevolutionary steps as completely new genetic systems.

Goldschmidt's views were largely rejected by the biological community. While he called homeotically mutant flies 'hopeful monsters' that might prosper, perhaps under some unusual environmental conditions, the prevailing view was that the homeotics were instead 'hopeless monsters'. They were doomed to being removed from natural populations whenever they occurred, because they inevitably reduced fitness. Fisher's theoretical approach strongly suggested this; and biologists who studied the homeotically transformed flies in the laboratory noted a drastic reduction in fitness that did not seem to be modifiable into a fitness increase by any changes in environmental conditions.

The second half of the twentieth century thus began with a prevailing view in evolutionary biology that saltationism was dead and gradualism had won the day. And this is still the prevailing view. We will return later to the question of whether this view is correct, or whether a more nuanced view of the relationship between gradual and sudden change in the evolution of development might be necessary.

The strong association (albeit not a universal one) between those who studied evolution from a developmental angle in the early twentieth century and saltationism is hardly an accident. If evolution works via gradual variation, the developmental variants involved in the evolutionary process are so subtle, and so similar to each other, that there is little to study from a developmental point of view. In contrast, if evolution works via large-effect mutations altering the developmental process, such as the homeotics, then there most certainly *is* something to study.

Having said that, the final twentieth-century figure whose work I will mention in this section – the British biologist Conrad Hal Waddington – had a rather different view of the role of development in evolution. He is mostly remembered for his publications on a phenomenon called genetic assimilation. Waddington showed that an unusual pattern of wing venation in *Drosophila* fruit-flies could be produced by subjecting metamorphosing pupae to a heat shock. This pattern (called 'crossveinless') was not universally found in the shocked flies, but it was quite common. When Waddington selectively bred from crossveinless flies, he found that in later generations of this stock that were *not* subjected to heat shock, the crossveinless pattern sometimes occurred spontaneously. This looks like a Lamarckian phenomenon – inheritance of an acquired character – but it is not. Waddington had been selecting for flies that had a higher probability of showing this pattern of veins, and hence for genes that conferred this higher probability. So an apparently Lamarckian process was actually a Darwinian one. Waddington emphasized the importance of the selective process just described, and also of another – selection to stabilize the development of the pattern that was 'fitter' in the context of his experiment (crossveinlessness is normally less fit rather than more so).

Although Waddington was not a neo-Lamarckian, he was critical of certain aspects of neo-Darwinism. Given that the saltationists, most notably Goldschmidt, had also been critical of neo-Darwinism, by the time evo-devo started in the 1980s many neo-Darwinians viewed developmental evolutionists with some suspicion. It is important to realize this fact, because it impinges on the philosophical outlooks taken by those who now see themselves as students of evo-devo (which are somewhat heterogeneous), and also on the outlooks of those present-day evolutionary biologists who are not students of evo-devo, but rather look at it from the outside – notably population geneticists.

Take-Home Messages From History

Here is a famous quote from the Spanish-American philosopher George Santayana: 'Those who cannot remember the past are condemned to repeat it.' Santayana made this statement in 1905, in Volume 1 of his philosophical treatise *The Life of Reason*, and there is much truth in it. In the context of doing science, it can be taken to mean that those scientists who are unaware of the history of their discipline – and of related ones – may unwittingly waste their time revisiting old debates that have either been resolved or abandoned as futile. With this point in mind, it is useful here to summarize the main messages that can be extracted from studies of the relationship between evolution and development in the period between 1800 and 1980.

(a) No Universal Laws in Biology

First, as in other fields of biology, the attempted construction of laws about evolution, development, or their interaction is not a sensible strategy. There are always too many exceptions. However, having said that, science is all about providing general explanations for ranges of processes and phenomena, not piecemeal explanations for specific ones. So we should endeavour to be as broad as possible in our generalizations. Sometimes, the limits of a general explanation of some evolutionary phenomenon correspond to the limits of a particular branch of the evolutionary tree of life. The British palaeontologist Alec Panchen called such limited generalizations 'taxonomic statements' in his 1992 book *Classification, Evolution and the Nature of Biology*. However, sometimes the limits of generalization do not map neatly to a particular branch of the evolutionary tree. As we saw in Chapter 1, the maximum potential span of an evo-devo generalization is the realm of multicellular creatures, which consists of a collection of several branches, some large and some small.

(b) A Messy but Interesting Relationship

Second, and following on from the above, we do not today accept the 'laws' of either von Baer or Haeckel. However, we do accept that there is some 'messy but interesting' probabilistic relationship between the course that the evolutionary lineage of a particular present-day species has taken over a long period of time and the course that the development of an individual belonging to that species takes over a much more limited period. For example, it seems unlikely that human embryos would have gill clefts if humans had not had fishy ancestors.

More generally, the human developmental trajectory encompasses stages that are a single cell (the zygote), a hollow quasi-spherical ball of cells (the blastula), a more complex form that has three tissue layers and is bilaterally symmetrical about an anteroposterior axis (the gastrula, followed by the neurula), and a later, larger elaboration of this (the pharyngula) that is characterized by those aforementioned gill clefts. These embryonic stages roughly parallel stages of our evolutionary ancestry. If we were able to examine the progress of our lineage from the original protocell called LUCA (the Last Universal Common Ancestor of all life on Earth) to present-day *Homo sapiens* more completely than the fossil record allows, we would probably see, in temporal sequence, the following stages: a single cell, a quasi-spherical assemblage of cells, a small simple bilaterally symmetrical animal without true organs (a sort of marine flatworm), and a larger, more elaborate bilaterally symmetrical animal with various organs, including gills (a sort of fish). But remember that we humans have direct development – there is no larval stage in our life cycle. For most animals that do have such a stage, whether aquatic (e.g. sea-urchins) or terrestrial (e.g. butterflies), the parallel between the developmental sequence of an individual and the evolutionary history of the species to which that individual belongs is more complex.

The fact that the relationship between evolution and development is messy should not be taken to downgrade it from the perspective of its scientific interest. Everything related to evolution is messy. It is, as we have already seen, a process that involves much historical contingency. And this contingency doesn't only come in the form of occasional asteroid impacts; rather, it is omnipresent. Even the relationship between Darwinian fitness and evolution is messy. Although graphs based on mathematical models of the spread of a new, fitter variant in a population often show an apparently smooth trend of increasing frequency of the new variant over time, most fitter variants disappear due to the random process of genetic drift long before natural selection gets a chance to act on them. This was emphasized by the British population geneticist J. B. S. Haldane in his 1932 book *The Causes of Evolution*.

(c) A Recurring Debate

Charles Darwin persuaded most biologists that evolution had happened. In terms of its main driving mechanism at the level of populations, he also (eventually) persuaded most biologists that this was natural selection. There have been no credible challenges to the former; the claims of the so-called 'creation scientists' are not to be taken seriously. Neither have there been any credible challenges to the latter – the efficacy and general importance of natural selection. However, selection can work with any type of heritable variation. In particular, it can operate regardless of whether the variation is continuous or discrete at the level of the phenotype. Where variation is discrete, the variants can differ in minor ways (e.g. blue vs. brown eyes in humans, each of which categories is also subject to continuous variation in shade), more major ways (e.g. dextral vs. sinistral snails, in which not just the shell but the body too shows a reversed direction of coiling), or very major ways (e.g. four-winged vs. two-winged fruit-flies).

Darwin tried to persuade us that all selectively driven evolutionary change was based on continuous rather than discrete variation, with his insistence on *natura non facit saltum*, as in his comment that 'natural selection can act only by taking advantage of slight successive variations; she can never take a leap' (Chapter 6 of *The Origin of Species*). Was he equally successful in this respect? Before answering this question, we should acknowledge that Darwin was very much aware of the existence of discrete variants; as noted earlier, he referred to some of these as 'sports'. But he considered them as unimportant to evolution. He saw them as a mixture of unfit and selectively neutral variants. In his view, none of them was actively favoured by natural selection.

Despite Darwin's insistence on *natura non facit saltum*, objections to his 'pan-gradualism' view of evolution have been a recurring theme from 1859 onwards. Darwin's bulldog, Thomas Henry Huxley, wrote to his friend on 23 November 1859, just after reading the 'hot off the press' *Origin of Species*, saying that 'you have loaded yourself with an unnecessary difficulty in adopting *Natura non facit saltum* so unreservedly'. Thus one of the foremost early pro-Darwinians was sceptical of the pan-gradualist view. As we have seen, saltationism has repeatedly reared its head since then, with Bateson, de Vries, Thompson, Goldschmidt, and others. Although their views have not been accepted in mainstream evolutionary theory, some practitioners of evo-devo are more open-minded on this issue than are most population geneticists. It is interesting that one of the first evo-devo books – *Embryos, Genes, and Evolution*, by Rudolf Raff and Thomas Kaufman (1983) – had a dedication 'To Richard Goldschmidt 1878–1958', despite the fact that Goldschmidt is among the most extreme saltationists of all.

There is little point in rerunning the gradualist versus saltationist debates of the past; in particular, it is pointless to try to raise from the dead hypotheses based on mutations that clearly reduce fitness rather than increase it. However, perhaps the advent of evo-devo allows us to revisit the issue of the nature of the genetic and developmental variation upon which selection acts in a more enlightened way. Perhaps, instead of discussing the abstract *magnitude* of a mutation's effect on the phenotype, we can discuss the *timing* of its effect in the developmental process, and the *nature* of its effect. Indeed, these three things – magnitude, timing, and nature – must be interrelated. And the way in which they are related is of considerable interest.

Finally, while gene mutation is the ultimate source of the variation on which selection acts, most variation at the level of developmental processes is more subtle and complex than being the result of a single base-change in the stretch of DNA that makes up a particular gene. The ubiquitous continuous variation in the outcome of development that we see in normal distribution curves for the sizes and shapes of morphological characters (e.g. neck length, whether in humans or giraffes) has long been understood to be complex in its causality. It is only partly heritable, and the heritable component has a multi-genic basis. Waddington's genetic assimilation adds a further subtlety to the way in which variation manifests itself. His work has contributed to a strand of evo-devo that focuses on the concept of 'developmental reaction norms' – the way in which development is influenced jointly by genes and environment. We'll examine this evo-devo concept in Chapter 5. First, though, we should take a look at several important concepts that belong not specifically to evo-devo but to evolutionary and developmental biology more generally.

3 Evolutionary and Developmental Essentials

Evolutionary Pattern

Our starting point for discussion of evolutionary pattern is the word 'clade'. This was introduced by the British biologist Julian Huxley (grandson of Darwin's bulldog T. H. Huxley) in the 1940s. It means a taxonomic group of a particular kind: one that includes all the descendants of a particular ancestral species, and no others. This kind of group can also be called monophyletic. When the German taxonomist Willi Hennig founded the new approach to taxonomy that we now call cladistics, in the 1950s and 1960s, the idea of a clade was central. For those not familiar with cladistics, Hennig's main concern was that the evolutionary trees that were used through much of the literature of evolutionary biology confounded two things: closeness of ancestry and similarity in body form.

Although there is a relationship between these two, with more closely related animals being generally more similar to each other (e.g. human, chimp) than are more distantly related animals (e.g. human, sea-urchin), this is another of those messy probabilistic relationships that is by no means guaranteed in individual cases – due largely to the occurrence of convergent evolution. This process involves lineages that started off being characterized by very different body forms becoming more similar in this respect, due to natural selection producing common structures. For example, many of the birds that we call swifts, swallows, and martins are superficially similar in appearance, with their characteristic forked tails. In the case of swallows and martins this similarity in form is due to closeness of evolutionary relationship. However, the similarity between these birds and the swifts is due to convergent

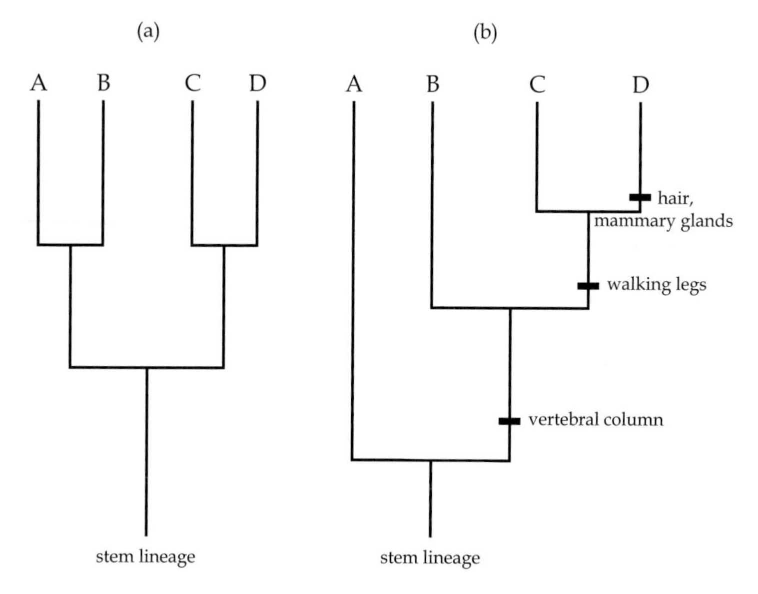

Figure 3.1 (a) Abstract cladogram showing the evolutionary relationships among four unspecified groups of animals (A to D). (b) Cladogram showing the pattern of lineage splitting leading to four groups of chordates, with horizontal bars indicating the points of origin of the named homologous features (here, A is a lancelet, B is a fish, C is a reptile, and D is a mammal).

evolution. Swifts are only very distantly related to the others. In fact, swifts are much more closely related to hummingbirds than they are to swallows and martins.

Given the problem of convergence – which is a common occurrence in evolution, not a rarity – Hennig argued that any evolutionary tree diagram attempting to show both genealogy and morphology simultaneously was likely to lead to confusion. He championed the approach of focusing on patterns of lineage divergence to the exclusion of the morphology of the animals concerned. He used very abstract-looking trees such as the one shown in Figure 3.1a, which are called cladograms. Such a diagram can be thought of as representing groups-within-groups. This approach was initially called phylogenetic systematics (by Hennig), but is now more often simply called cladistics; it is closely linked to the pre-existing concept of homology, which we will now examine – from a starting point of the structure of an arm.

The human arm has a skeletal basis consisting of a single long bone, the humerus, followed by two long bones running in parallel, the radius and ulna. We can ignore, in the present context, the bones that are proximal to these (the pectoral girdle) and those that are distal (the bones of the wrist and hand). Equivalent bones to the human humerus, radius, and ulna can be found in all other mammals, including tree-living forms (e.g. monkeys), ground-living forms (e.g. horses), flying or gliding forms (e.g. bats), and aquatic forms (e.g. whales). This similarity of structure derives not from convergent evolution (as in the case of the forked tails of swallows and swifts), but rather from common ancestry. Such inherited similarity of form was called homology by the nineteenth-century English anatomist Richard Owen – though bizarrely Owen was a leading anti-Darwinian of his era. Homology came to be regarded as one of the main types of evidence for evolution. After all, what engineer (or intelligent designer), setting out to design three sets of appendages – for positioning tool-using hands, for moving across land, and for propelling an animal through water – would adopt the same bone-pattern for all three?

Homology is a key concept for linking evolutionary biology, developmental biology, evo-devo, and taxonomy. Any large taxonomic group of animals – e.g. a phylum – can be seen as a series of nested homologies, corresponding to the evolutionary idea of groups-within-groups. Take, for example, our own phylum, the Chordata. Every chordate has a notochord at some stage of its life, so this is a pan-chordate homology. Members of our own sub-phylum – Vertebrata – share a homologous skeleton, while non-vertebrate chordates, such as lancelets and sea-squirts, do not. Fur is a mammalian homology – it is found to a greater or lesser degree in all mammals (minimal in whales and dolphins) but not in other sorts of vertebrate. Each order within the class Mammalia is characterized by a more or less obvious homology that it alone possesses – for example the unique teeth of rodents.

The origins of new features that later fan out into descendant lineages can be noted on a cladogram, thus beginning to put some details of morphology back into the initially form-free diagram, as shown in Figure 3.1b. However, most of the chordate homologies discussed so far, such as skeletons and fur, tend to be mentally pictured in the adult. Early embryos have neither skeletons nor fur, though they do have notochords. So it is now interesting to ask whether homology of adult (and juvenile) features goes hand-in-hand with homology of embryogenesis and of developmental genes – these, after all, are the things that generate the forms of later stages.

Adults with homologous structures usually (though not always) exhibit similar development, with the homologous features being recognizable from a certain stage in embryogenesis right through to the adult. When adult forms are convergent rather than homologous, their developmental trajectories are usually *not* similar. For example, there is a marsupial mole, which is only distantly related to the 'ordinary' (placental) mole, to which it is convergent. Given their separate origins from rather distantly related mammalian lineages, the development of each reflects its phylogenetic position rather than its adult body form. Marsupial moles have marsupial development, while placental moles have placental development.

Given the good – albeit not universal – correspondence between homology/convergence of adult structure and similar/different development, it seemed reasonable to expect that the third dichotomy of similar/different developmental genes would align itself with the other two. However, one of the biggest surprises to have come out of evo-devo research so far is that this is not the case. Rather, we find that homologous developmental genes are used in many cases of convergent structures (e.g. vertebrate and arthropod legs), as well as in cases of homologous structures. The reason for this is to be found in the concept of gene co-option, which we'll discuss in Chapter 8.

We now need to return to trees for a moment. Hennig's philosophy, both in general and as a rationale for the type of tree that is now called a cladogram, was 'pattern before process'. In other words, before trying to understand the evolutionary processes that transform one type of structure into another (e.g. a fin into a leg), we need to be clear about the pattern of lineage splitting that actually took place. In this he was absolutely right, because if we proceed in a different way we end up trying to explain putative evolutionary processes that never actually happened.

An example of this can be found in twentieth-century attempts to understand the supposed evolutionary process of 'arthropodization', in which a segmented worm (without an exoskeleton) was transformed into a segmented arthropod (with an exoskeleton). The desire to explain this process arose from a perception that, at the level of animal phyla, the segmented worms (annelids) were closely related to the arthropods. This was discovered to be untrue in 1997, with the publication of a key paper in *Nature* by Anna Marie Aguinaldo *et al.* What they discovered was that growth by moulting (or lack thereof) was a better morphological character than segmentation (or lack thereof) to use for determining the high-level structure of the animal kingdom, in terms of the relationships among phyla. However, these authors did not reach their conclusion via a detailed comparative study of the developmental biology of moulting. Rather, they studied comparative DNA sequence data; this was then linked to moulting in a post-hoc manner. And in general these days DNA sequence data are used in preference to morphological data for reconstructing trees, because although no source of data is without its problems, the problem that convergence poses for tree reconstruction is less serious with molecular than with morphological data.

Evolutionary Process

Having looked at some essentials in the area of evolutionary pattern, we turn now to evolutionary process. This takes us to Darwinian natural selection. One of the most important distinctions between different types of selection is that between directional and stabilizing. As their names suggest, these cause evolutionary change and stasis respectively. For much of evolutionary history, many characters in many lineages are subject to stabilizing selection – in other words, the fittest variants are those that are close to the average in the overall distribution of all variants. However, at any point in evolutionary time, some characters in some lineages are undergoing directional selection because the current mean phenotype and the fittest phenotype do not coincide, perhaps due to recent environmental change. Here, one tail of the distribution is favoured by selection, and so the whole phenotypic distribution shifts accordingly. (The above account is based on continuous variation; if the variation is instead discrete, then the same distinction exists between the two types of selection, though the stabilizing kind is then usually called balancing.)

In population genetics, there has been much interest in both directional and stabilizing/balancing selection. In evo-devo, evolutionary change in development is the main focus of attention. For example, we are generally more interested in how a dinosaur can evolve into a bird rather than in how a dinosaur can remain a dinosaur. So, directional selection is the more important of the two in this field, but with a qualification, as follows. There is much interest, in evo-devo, in various aspects of evolutionary constraint – we noted this already, and will delve into it further in Chapter 4. Stabilizing selection can also be called selective constraint – because it is a form of selection that constrains, rather than assists, evolutionary change. A von Baerian pattern of embryonic divergence may well be the result of the way the balance between stabilizing and directional selection changes between different developmental stages.

Let's now 'drill down' further into directional selection. There is an important distinction within this type of selection that is often overlooked: that between the kinds of directional selection producing *general* and *special* adaptation. This distinction may be of particular importance to evo devo.

The key issue here is the *environmental breadth* of an adaptation. Let's compare two very different adaptations to illustrate this point. In the case of an ancestral, or 'stem', cetacean (the whale/dolphin group) evolving a streamlined body shape, the environment to which the new body shape is adapted is very broad: aquatic environments prevail across more than 70% of the Earth's surface. In contrast, the environment to which the melanic form of the famous moth *Biston betularia* was adapted, namely soot-blackened, lichen-free surfaces, was very narrow. It was limited in space (urban areas versus their surrounding rural ones) and, as it turned out, also in time (the soot is now largely gone). Because of its restricted environmental breadth, the *Biston* pigmentation example can be thought of as a special adaptation. The cetacean body-shape example is clearly more general. More general again is the evolution of hinged jaws in an early jawless fish. In this case, the adaptation works in all environments, not just aquatic ones, and hence has been retained in virtually all descendants of the original gnathostome ('jawed-mouth'), including both terrestrial and aquatic ones. If there is any such thing as a *completely* general adaptation, this is an example of it. Ironically, though, vertebrate jaws may have initially evolved to help ventilation (as a sort of pumping mechanism), and may only have been redeployed later for feeding purposes.

The American biologist William L. Brown wrote an important paper on this issue of the breadth of adaptations in 1958. He distinguished special and general adaptations as follows:

> Both at and above species level, special adaptations appear to function to fit the organism or population to the particular features of the immediate environment. General adaptations are concerned more with internal organization and efficiency of the individuals or population, more or less independently of the local details of the environment.

Brown then emphasized the importance of general adaptation, and of recognizing its distinctness from special adaptation: 'general adaptation seems to me to represent one of the potentially more important factors in evolution'; and 'the distinction between general and special adaptation may well be a fundamental one, heretofore not widely enough considered in evolutionary thinking'. In my view, Brown was right at the time he wrote, and his point remains valid today.

Brown approached this whole topic from the perspective of ecology. However, his approach is very helpful also from an evo-devo perspective. What Brown calls general adaptation probably often takes the form of improved *coadaptation* – the adaptedness of different parts of an animal to each other. This could relate to adult function – for example after the origin of hinged jaws in vertebrates there was probably strong selection for improved articulation between the upper and lower jaws. It could also relate to development – for example some change in the early embryogenesis of a mammal that enhances the probability of a smooth transition to later stages.

Although Brown's distinction between special and general adaptation, as given above, makes it seem like a dichotomy, he goes on to clarify that he is not arguing for two neatly defined categories of adaptation, without any intermediates: 'Of course, the two types of adaptation differ as extremes of emphasis, and not as absolutes.' One way of interpreting this statement is that there is a continuum in the environmental breadth of an adaptation, with 'special' at one end, 'general' at the other, and 'intermediate' in the middle.

A useful way to approach the continuum from special to general adaptation is to use trans-environment fitness profiles. Figure 3.2 shows eight examples of such profiles for the simple case of there being just two environments. In each panel, the line shows the relative fitness of a new variant compared to an

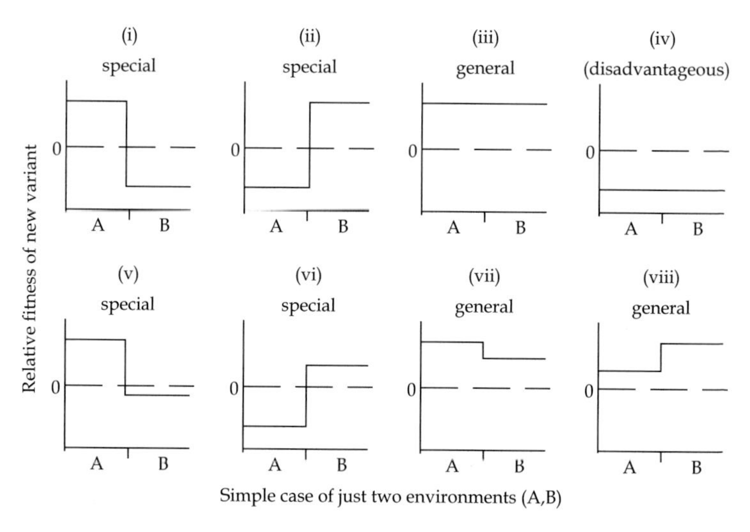

Figure 3.2 Illustrations of the difference between special and general adaptation, in the simplified case of just two environments (A and B). In each case the solid line shows the fitness of a new variant in the population relative to the original one (with zero on the vertical axis representing equal fitness). Top: fitness differential variable in direction but not in magnitude, so the 'steps' are always the same size. Bottom: fitness differential variable in both direction and magnitude, so any size of step is possible. For commentary on individual situations (i) to (viii), see text.

existing one, and how this changes between environments A and B, which differ in some (unspecified) way. The top four panels refer to the restricted case of a fitness difference between the two variants that can change in direction but not in degree. This is unrealistic, but it is a good starting point. With this restriction, there are only four possible trans-environment fitness profiles. These represent: (i) new variant advantageous in environment A but disadvantageous in B; (ii) new variant advantageous in B but disadvantageous in A; (iii) new variant advantageous in both; (iv) new variant disadvantageous in both. The first two of these correspond to different special adaptations; the third corresponds to general adaptation; the last is of little interest, as the new variant is selected against in both environments and so there will be no evolutionary change. The bottom panels show four examples (out of the infinite number of possibilities that exist) when the relative fitness difference between the new and original variants is variable in degree as well as in direction. These represent: (v) and (vi) special adaptation; (vii) and (viii) general adaptation.

There has been much emphasis throughout the history of evolutionary biology on the fact that natural selection is a process that is very much restricted to the moment. It has no ability to look ahead in time to consider what direction of evolutionary change might be beneficial later on – even just a few generations later on. So, natural selection has often been described as 'blind'. An early example of such usage can be found in the 1942 book *Evolution: The Modern Synthesis*, by Julian Huxley, who says that selection is 'blind and mechanical'. More recently (and more famously), Richard Dawkins described natural selection as being a 'blind, unconscious, automatic process', and he followed this up by saying that selection has 'no foresight, no sight at all'; these quotes are from *The Blind Watchmaker*, published in 1986. The emphasis in most discussions of this subject, including both Huxley's and Dawkins', is very much on the lack of foresight, in other words on the lack of natural selection's ability to see forward *in time*. However, in looking at the environmental breadth of an adaptation, we are emphasizing natural selection's inability to see beyond the current location *in space*. For this, instead of using Dawkins' term 'blind watchmaker', we might use the term 'blinkered watchmaker', taken from the blinkers that can be attached to the bridle of a working horse to prevent it from being distracted by things that are happening to its side.

As evolution proceeds, over millions of generations, and each species with general adaptations spreads over many new areas, it may 'bud off' descendant species with special adaptations and more restricted ranges. This is an important idea, because it provides one hypothesis to explain the uneven distribution of actual animal forms across the 'theoretical morphospace' of all possible forms. The alternative hypothesis is that of developmental constraint (see next chapter).

But what exactly is 'theoretical morphospace'? This is the first time I have mentioned this important concept, so I should explain it. Imagine a three-dimensional graph – a transparent cube whose axes are three measurable characteristics of animals, for example body length, number of eyes, and number of pairs of appendages. Every point within the cube represents a theoretical animal. But actual animals are not randomly distributed throughout the cube. Rather, they are clustered in certain regions of the morphospace that the cube represents. For example, there are many animals with four legs and many with six, very few with 150 legs, and none with 2000 legs. Those with lots of legs are generally small – because no vertebrates are multi-legged – though there were giant millipedes in the Palaeozoic era. There are many animals with two eyes, some with four, some with eight, and some with other numbers. As with legs, there is an interaction with body size: animals with many eyes, like those with many legs, are usually small; they are often spiders. The more dimensions (body characteristics) we build into the picture, the more apparent the non-random distribution of actual animal forms becomes.

Whether this pattern of clumped occupation of morphospace is explicable through the hypothesis of an interplay between general and special adaptation or through the hypothesis of developmental constraint (or through both) is a major unresolved issue.

Development in the Context of Life Cycles

In humans, the relationship between development and the life cycle is reasonably clear. (Recall that we have what is called direct development, meaning that there is no larva.) There is a fairly well-defined developmental phase in the life cycle, lasting from the zygote to the onset of reproductive maturity, and hence by definition (in biology but not in society) to adulthood. This phase can be subdivided into embryogenesis and post-embryonic development. Growth predominates in the final stages of the former and for much of the latter. However, note that human post-embryonic developmental processes include some that cannot be described just as growth. The eruption of the milk teeth takes place over the first two years or so of life; and their replacement with the so-called permanent teeth occurs in later development, extending over a much longer period of years. And of course there is the development of sexual characteristics at puberty.

Although the changes occurring at puberty effectively end the developmental period of the life cycle, the wisdom teeth are yet to develop, so the threshold between the two phases is not clear-cut. There are other post-puberty developmental processes in humans too, though exactly how many depends on how development is defined. The process of wound healing can certainly be described as developmental; so can the replacement of short-lived cells of the stomach lining with new ones produced from nearby stem cells. Other vertebrates exhibit the developmental phenomenon of regeneration: for example, some species of lizards can regenerate an excised tail. Many invertebrates exhibit an even greater capacity for regeneration. Earthworms can regenerate considerable stretches of their bodies following excision, and some can even regenerate at both ends following bisection. However, it must be emphasized that this capacity varies considerably among species of earthworms.

Now we turn to focus on those life cycles that are characterized by indirect development, including polychaete worms (marine relatives of earthworms), sea-snails, butterflies, and frogs. These all have two main developmental phases that involve much more than growth – embryogenesis (producing a larva) and metamorphosis (producing an adult). But other animals have even more complex life cycles. Many of these are parasitic, and their life cycle complexity is connected with their hosts, with different stages often parasitizing host species from different phyla. Flukes are parasitic animals belonging to the flatworm phylum Platyhelminthes. Those of the genus *Schistosoma* cause the disease called bilharzia (or schistosomiasis) in humans. But at another stage of their life cycle these flukes parasitize snails. The complete life cycle of *Schistosoma* has eight stages.

We noted above that even in simple life cycles with direct development, such as those of humans and other mammals, the distinction between the different phases of the cycle is not clear-cut. It can be argued that in complex life cycles it is even less so. In the following paragraph I give the butterfly life cycle in a little more detail than the usual shorthand version of egg → caterpillar → chrysalis → adult, to illustrate this point.

Embryogenesis in a butterfly egg ends with the hatching of a baby caterpillar. Within its body are small structures called imaginal discs that will later (at metamorphosis) be used to build the adult body (an adult insect is technically known as an imago). The caterpillar grows by a series of moults, the number of which varies among species. Eventually, the large caterpillar becomes a chrysalis – a generally immobile entity with a hard protective exterior. Inside, most of the caterpillar's tissues are broken down. The cells of the imaginal discs proliferate and form the various adult structures. Eventually, the casing of the chrysalis breaks open and the butterfly emerges. If it survives for long enough it will mate, and the resultant eggs start the cycle anew.

We can now see that during the caterpillar phase of the life cycle two different developmental processes are occurring in parallel: the growth of the caterpillar and the quasi-independent growth within it of the little discs of tissue from which the butterfly will be developed later. However, once the adult stage is reached, this has few processes that could be called developmental ones; butterflies cannot regenerate lost wings or legs.

How we look at the overall developmental process, in the cases of both simple and complex life cycles, is of some philosophical interest. As adult humans, we naturally tend to take what has been criticized as an 'adultocentric view' by the Italian biologist Alessandro Minelli. And perhaps adult *biologists* have even more of a tendency to take such a view. Minelli puts it as follows in his 2009 book *Perspectives in Animal Phylogeny and Evolution*:

> The adult is, by definition, the reproductive stage and Darwinian fitness is measured in terms of an animal's contribution to the gene pool in the next generation. This requires reproduction, and the measure of Darwinian fitness is based on the performance of the adult. This seems to justify seeing embryos and juveniles simply as preparatory stages eventually leading to the all-important reproductive stage.

Minelli is right (a) that selective scenarios are often pictured in terms of the fitness of adults, and (b) that those who have an automatic tendency to see them thus should mend their ways. However, it is clear that while many biologists have indeed been guilty of adultocentrism, Darwin was not one of them. He emphasized that fitness differences apply at all stages of the life cycle. In an 1857 letter to Asa Gray, which was published as part of the famous Darwin–Wallace paper of 1858, he says that 'an organic being' may over a period of time 'come to be adapted to a score of contingencies – natural selection accumulating those slight variations in all parts of its structure, which are in any way useful to it *during any part of its life*' (the italics are mine). So, while the adultocentrism that characterizes many publications in evolutionary biology between then and now may have its origins in a particular way of thinking about natural selection, it does not have its origins in Darwinism as originally formulated.

Model Organisms

The living world can be divided up in many ways. From the perspective of how animals develop, the division between simple life cycles (direct development) and complex life cycles (indirect development) is one of them. But from the perspective of *studying* development, a very different form of division is into model and non-model organisms. The former group is very small compared to the latter but has made a disproportionate contribution to the discipline of developmental biology. Thus it's important to understand this frequently encountered division.

A model organism in developmental biology is a species that got chosen as a workhorse for studies in this field because of certain features of its life cycle that make it particularly amenable to study in comparison with most other species. The domestic chicken *Gallus gallus* was a model organism from an early point in the history of the discipline, indeed from long before the term 'model organism' was invented. This was due partly to its ready availability and partly to its large external eggs. Experimenters interested in the development of the chick can, with care, make 'windows' in the shell and watch development happening. Imagine trying to create such a window in the development of a placental mammal.

Amphibians, and in particular the African clawed frog *Xenopus laevis*, also became popular model organisms for several reasons, with one of them again being ease of study of the early developmental processes that take place in the egg. A typical frog's egg is even more amenable to study than a hen's egg, since it is transparent.

By the time of the discovery of the homeobox in the early 1980s, the fruit-fly *Drosophila melanogaster* had become a well-established model organism not just for genetics (which it had been for decades) but also for developmental biology. The latter followed from the former. The usefulness of *Drosophila* for developmental biology research stemmed largely from all the available genetic knowledge and a wide range of well-practised genetic techniques that had been built up over the years. Indeed, *Drosophila* was instrumental in the establishment of the combined field of developmental genetics, which preceded the establishment of evo-devo.

The millimetre-long nematode worm *Caenorhabditis elegans* was established as a model system by the South African biologist Sydney Brenner in the 1970s.

This worm has a life cycle that is even shorter than that of *Drosophila melanogaster* – a few days in the laboratory, rather than a couple of weeks. It also turns out to have a very fixed cell lineage – the pattern of cell division leading from zygote to adult. And at the end of development it has a fixed number of cells – exactly 959 of them in the body (soma) of a hermaphroditic adult.

The main model systems for developmental biology are often said to be seven in number – the above four plus the house mouse *Mus musculus*, the freshwater zebrafish *Danio rerio*, and, in the plant kingdom, the thale cress *Arabidopsis thaliana*. Thus the number of non-model species is considerably more than a million, regardless of whether we are talking about the animal kingdom or 'animals + plants', or all multicellular life forms. So this is clearly a very asymmetrical division of the living world. It is also a less clear-cut one than I have painted it so far. Some species are at the other end of the spectrum from the seven model systems listed above – for example orang-utans, elephants, lions, komodo dragons, giant squid, deep-sea tube-worms, and maple trees, to name seven very definitely non-model organisms. This second list of seven helps to emphasize certain features that model organisms must lack but that many non-model organisms possess, including large adult body size, long generation-time, fierceness, environmental inaccessibility, endangered conservation status, and serious ethical issues in terms of experimentation.

In between our two lists of seven, there are countless species with very varied utility for conducting developmental biology research. Species that can be regarded as quasi-model systems for various reasons include the following: newts of the genus *Triturus* (used by early embryologists, notably Hans Spemann and Hilde Mangold), species of the cnidarian genus *Hydra* (used in the study of regeneration), and sea-urchins of the genus *Strongylocentrotus* (used by the American biologist Rudolf Raff and his colleagues to study the evolution of feeding and non-feeding types of larvae). Many others could be added to this list, including some that could be regarded as incipient model organisms that may come to join the 'big seven' in the future.

The importance of model organisms for developmental biology is hard to overstate. By concentrating on these few creatures we have learned so much more than we would have by concentrating on seven randomly chosen ones, or by spreading research evenly across a much greater number of species.

Furthermore, because developmental processes and the genes that are involved in them have turned out to be much less species-specific than was originally thought, the discoveries that have emerged from the development of model organisms are much more widely applicable than the early biologists could have imagined, even in their wildest dreams. For example the British biologist Norman Berrill said in his 1961 book *Growth, Development, and Pattern*, that 'no general theory of development has emerged, in spite of the mounting mass of observational and experimental information'. Although the genetic toolkit of animal development that was a major evo-devo finding (see next chapter) is not a theory *per se*, it has helped to usher in a new era in developmental biology where the idea of generalizing about the development of morphologically very different animals looks much less crazy than it did in the past.

The Nature of the Developmental Process

What is the essential nature of this thing we call 'development'? For an extended exploration of this question, I would direct the reader to the companion book in this series, *Understanding Development*, by Alessandro Minelli. Here, I will take a more focused approach. For example, I do not intend to discuss questions such as whether tumour formation, aging, and metabolism should be included in a broad view of development. Rather, I want to restrict attention to the 'main' processes of embryogenesis and post-embryonic development – the latter including, where appropriate, metamorphosis. This is already a sufficiently great challenge without extending it further into the realms of pathology, senescence, or biochemistry.

At an abstract level, development is a network of causal links, with an individual link being A → B, where A and B are any two entities involved in the developmental process (e.g. developmental gene makes developmentally active protein). When there are multiple interconnected causal links, the simplest arrangement is one in which they are all connected up in a line or series, for example A → B → C → D. This can be referred to as a developmental pathway or a developmental cascade, though usually when biologists use these terms they know that in reality the situation is rarely if ever this simple. Typically, there are divergences and/or convergences between pathways. Where two developmental entities A and B are both required to cause

C, this is called a combinatorial system; if, in contrast, A causes B and C, this can be thought of as part of a branching developmental hierarchy.

The overall developmental process in any particular type of animal is so complex that it is probably best described as a *network* in which some components are quasi-cascades, some are combinatorial, and some are hierarchical. In addition, it will include components that are not explicitly covered by any of those three terms, for example negative feedback loops, which act like thermostats to regulate a developmental process. Even when development is described in this simplified abstract way, in terms of unspecified causal links, we can see that it is a very complex process. However, when we inject levels of spatial scale into the picture, it becomes more complex still.

To understand development, we need to be able to interconnect processes that are happening at the molecular level (e.g. a gene), the cellular level (e.g. a stem cell that is differentiating into a blood cell), and the tissue level. An example of the tissue level would be a sheet of hundreds of cells in which cell differentiation is happening in a very specific way in space and time (pattern formation), while, simultaneously, the sheet is forming itself into a different shape, such as a tube (morphogenesis). Having an understanding of processes at one level does not necessarily help much in understanding processes at another. For example, even if I understood exactly how some cells in my hand become muscle cells, that knowledge might not help me to understand why my fingers end up a different shape from my thumb (and, to a lesser extent, from each other). Let's now look at these three levels of organization and the interactions between them.

Here are some important things that happen at the molecular/cellular level in the context of cell–cell signalling, which we first came across in the context of the Hedgehog protein in Chapter 1. For simplicity, we can imagine these things happening strictly in series (Figure 3.3). A developmental gene in one cell makes RNA and hence a protein (1), which is secreted out of the cell (2), attaches to receptors on the outside of another cell (3), and in so doing switches on another gene inside that cell. This 'switching on from the outside' works via a series of transmembrane (4), cytoplasmic (5), and into-the-nucleus (6) causal links. After the sixth step in this series, the protein that has moved into the cell nucleus, which is known as a transcription factor, binds to the

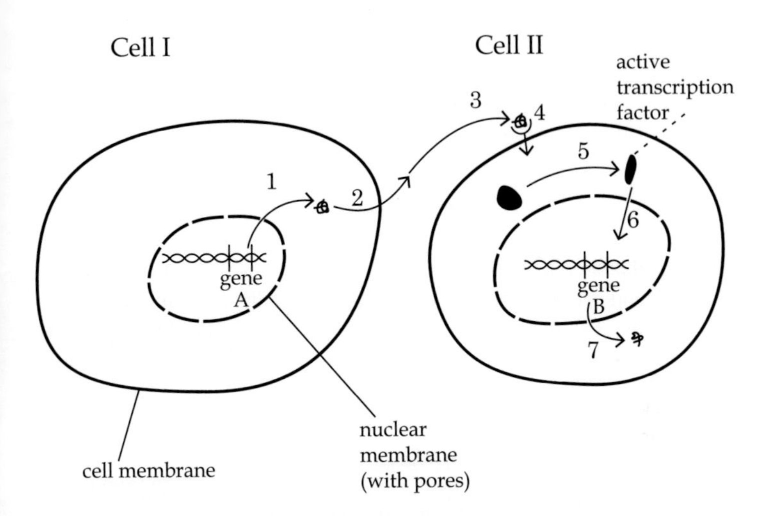

Figure 3.3 A generalized cell–cell signalling pathway. A gene in one cell makes a protein that is secreted out of that cell and received by another cell, where its receipt causes a target gene to be switched on. This process takes place via a series of steps, 1 through 7, which are explained in the text.

DNA of a target gene. This binding initiates the transcription of RNA from this gene, which is then translated into its corresponding protein (7). The transcription factor was made previously by another developmental gene (not shown in the figure), but it was in a dormant state prior to its activation.

The process described above probably sounds complex to anyone unfamiliar with developmental biology or genetics. And yet it is a very simplified version of reality – the developmental reality of a constant flux of multiple players of many kinds interacting in both binary (on/off) and quantitative (varying in degree) ways. Moreover, the above account was very limited in both space and time. Spatially, we considered only two cells out of many. Recall from our tour of model organisms that 'many' can mean about 1000 (for a nematode worm) or many millions (for a mammal). Temporally, the process as described could have taken place in a couple of minutes, out of the many minutes that there are in a life cycle. Recall that animal life cycles can be measured in days, weeks, months, or years, depending on the species concerned.

At the level of an individual gene within a cell that receives a signal from outside, the result is often 'being switched on', as described above. But sometimes incoming signals provoke a whole-cell response. A whole cell can do four main things: divide, move, differentiate, or die. The form of an adult animal is the result of the ways in which these four cellular activities are controlled in time and space, and the ways in which they interact with each other.

Cell division predominates in the earliest stages of animal development when the zygote undergoes the process of cleavage – literally, the cell cleaving into two, then four, then eight, etc. Following on immediately after cleavage is the process of gastrulation ('stomach-making'), in which the embryo becomes three-layered. The outer layer (ectoderm) will eventually form the skin and central nervous system, the middle one (mesoderm) the skeletal musculature of the body, and the inner one (endoderm) the gut. As well as becoming layered at this stage of development, the embryo alters its shape from quasi-spherical to quasi-sausage. It comes to have a head-end and a tail-end. It becomes bilaterally symmetrical. The gut is formed by migration of cells into its interior, producing what can be called 'a tube within a tube'.

The extending internal tube eventually makes connection with the outside world by forming a hole at the other end of the embryo from the end at which it started. In one large group of bilaterian animals (protostomes), the first hole typically becomes the mouth, the latter the anus; in the other large group (deuterostomes), the first hole typically becomes the anus, the second one the mouth. Of course, since evolution is a messy business, there are exceptions to this general rule; nevertheless, the large groups that we call 'mouth-first' and 'mouth-second' (Greek: *stoma* = mouth, *proto* = first, *deutero* = second) provide a useful way of recognizing the relationships between major groups of animals, such as phyla. For example, this distinction was used a long time ago to claim a close relationship between two seemingly very different groups – the chordates and the echinoderms. Molecular data later confirmed that we and the likes of sea-urchins are indeed close evolutionary 'cousins'.

So the first phase of development – cleavage – is dominated by cell division. While division continues into the second – gastrulation – this phase also involves much cell movement. The third phase is called neurulation (or neurogenesis, depending on the taxon), as it sees the early development of the central nervous system. In a vertebrate, neurulation occurs on the dorsal side of the embryo. Recall that, in an arthropod, neurogenesis occurs on the ventral side, but the following account will be based on vertebrates. Neurulation is characterized by more elaborate cell differentiation than occurred in the stages before it, as we will now see.

Along the dorsal midline, the neural tube that will later become our brain and spinal cord forms. A cross-section of the embryo at this stage shows that this tube is overlain by a sheet of ectoderm and underlain by the notochord (Figure 3.4). At first, the cells of the neural tube are all roughly equivalent. However, they soon become differentiated into various cell types. This happens under the influence of developmental signals emanating from the overlying and underlying tissue. For simplicity, we will just consider the signals coming from below – in other words from the notochord.

Cells of the notochord secrete a signalling protein belonging to the Hedgehog family, which we met earlier in the context of the class-naming *hedgehog* gene in fruit-flies. In vertebrates, there are several Hedgehog-family genes. The one that is important in signalling from the notochord to the neural tube is called *Sonic hedgehog* – or *Shh* for short. Not only do notochord cells have this gene switched on, but so too, after they receive the signal, do cells of the ventral-most region of the tube, called the floor-plate (Figure 3.4). So this region of the tube comes to have the highest concentration of Shh protein. The concentration drops towards the dorsal part of the tube. Cells in the tube respond to certain thresholds in the concentration of Shh; they consequently differentiate as different types of nerve cell, with some becoming the motor neurons that control our muscles (and that die, with disastrous consequences, in motor neuron disease).

So, neurulation could be described as a process dominated by cell differentiation. However, the earlier-mentioned processes of cell division and cell movement are also still taking place. The neural tube was itself produced by these processes – a flat neural plate buckling up at its edges and the upward flaps of tissue meeting in the middle and joining together. Some cells – called neural crest cells – detach from the dorsal part of the tube and migrate to various parts of the body, where they have various fates, in part determined by their initial position along the anteroposterior body axis.

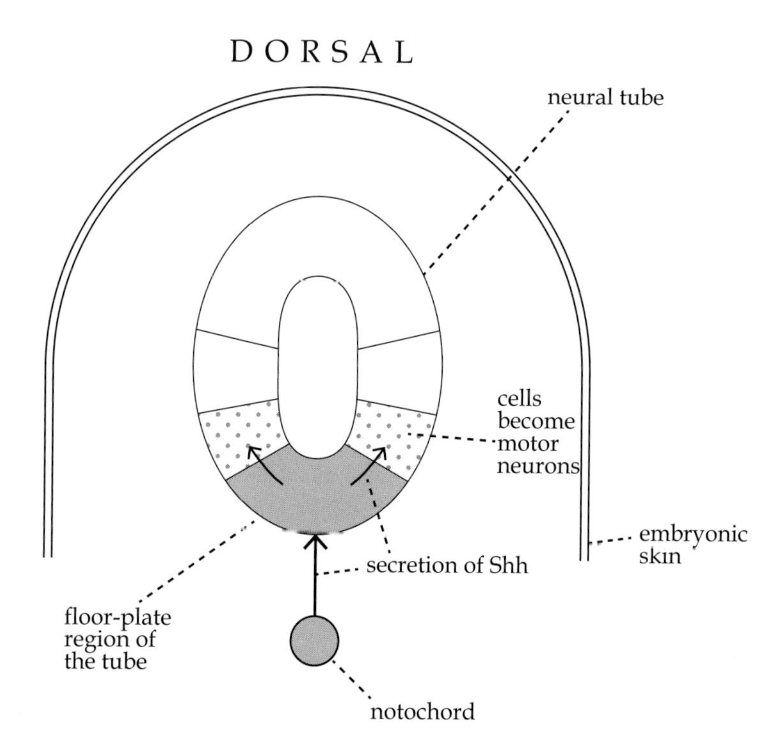

Figure 3.4 Cell differentiation in the developing neural tube of a vertebrate embryo. The protein called Sonic hedgehog (Shh) is secreted by cells of the notochord and, later, by cells of the floor-plate region of the tube. As a result, a dorsoventral concentration gradient of this protein is established. Cells in the lateral wall of the tube respond to this gradient by differentiating in various ways. For example, some cells that are exposed to high concentrations become motor neurons.

Of the four things that cells do, we have now seen examples of three, but not yet the fourth – cell death. Although I mentioned this in relation to motor neuron disease, that involves *pathological* cell death. In development, a different form of death is important – *programmed* cell death, or apoptosis. The development of our hands provides an example of this process. In the limb buds of a human embryo, flaps of tissue extend between the digits – fingers or toes. However, as neonates we lack such tissue; the reason is that apoptosis has occurred in the interdigital areas. In a species of mammal or bird with webbed feet, such as a platypus or a duck, there is less apoptosis in these regions.

The enormous complexity of development should now be clear. And the complexity of evolution is self-evident. Given these facts, you might be forgiven for thinking that evo-devo is a doomed endeavour. After all, surely the interaction between two complex processes must be more complex again? Well, yes and no. Strangely, the advent of evo-devo has helped us to understand the nature of both evolution and development better than we did before the 1980s, as well as having allowed us to make a start on understanding how they interact with each other. These improvements in understanding have involved concepts that can be thought of as belonging to evo-devo rather than the flanking disciplines. These are the 'essentials' that we will discuss in the next chapter.

4 Evo-Devo Essentials

Developmental Repatterning

Although I received my doctoral training within the neo-Darwinian tradition, in a university department (at Nottingham) that was largely devoted to population genetics, there is a view of evolution adopted by some neo-Darwinians that I have always rebelled against. This is the view that evolutionary processes can be understood in terms of only two levels of biological organization – the gene and the population. At its worst, this view is associated with actually defining evolution in those terms alone. For example, in their 1971 book *A Primer of Population Biology*, the American biologists Edward O. Wilson and William H. Bossert defined evolution as 'a change in the gene frequency of a population'. Evo-devo can be seen as a rebellion against this overly reductionist approach.

The term 'gene frequency' applies to the situation in which there are (at least) two versions of a gene (alleles) in a given population of a given species – say the population of great ramshorn snails in a large pond – at a given time. Assuming that one version of the gene was the original one, the other(s) have arisen from it by mutation. But mutation is just a change in the gene itself. From an organism's point of view, what is important is the resultant change in what the product of the gene does. In the case of developmental genes, the effect of such a mutation can lead to a wide range of possible outcomes, for which there was until recently only a very partial terminology. Now it is closer to being complete, though it will probably need to evolve further in the next few years.

A cover-term that can be used for all mutational and evolutionary changes in development is 'developmental repatterning'. This means a change in something that is itself a process of change. Although geneticists often talk about a 'phenotypic' change – where a certain phenotype could be the eye colour or the wing structure of a fruit-fly – this is 3D shorthand for a 4D phenomenon, namely the repatterning of development. Note that, without further elaboration, an instance of developmental repatterning could take any form at all; it could involve alteration of any of the developmental processes that we looked at in the previous chapter, or any combination of them. The term 'developmental repatterning' is a fusion of Gerhard Roth and David Wake's 1985 'ontogenetic repatterning' and my own 'developmental reprogramming'. I do not use the former label here because it was used by those authors to include most, but not all, sorts of evolutionary change in development; and I do not use my own earlier term (I abandoned it about a decade ago) because 'reprogramming' has undesirable philosophical implications.

A useful way to classify types of repatterning, and one that I began to use in 2011 in my book *Evolution: A Developmental Approach*, is into changes in time, space, amount, or type; these are called, respectively, heterochrony, heterotopy, heterometry, and heterotypy. They can be recognized at different levels, from that of a single developmental gene to that of the whole organism. At the level of a developmental gene, mutation in the coding part of the gene will lead to a slightly different type of protein product (heterotypy); in contrast, mutation in the regulatory region of the gene, to which various controlling factors attach in order to switch the gene on, may cause the gene to be expressed (switched on) at different times (heterochrony), in different places (heterotopy), or at different rates (heterometry).

These four types of developmental repatterning can also be recognized at the level of the whole organism and its development. The salamander species *Ambystoma mexicanum*, commonly known as the axolotl, has undergone a kind of heterochronic change that has led to the aquatic stage (normally just a pre-reproductive larva) becoming reproductively mature. In other members of its genus, such as the tiger salamander *Ambystoma tigrinum*, the normal situation is that reproductive maturity is not reached until metamorphosis to a land-dwelling adult. This type of heterochronic change – slowing of somatic development relative to reproductive development – is called neoteny.

An example of heterotopy at the level of the organism is the evolution of the centipede species *Henia vesuviana*. In most species of centipede, the sacs containing the venom that is injected into prey animals by the venom claws are located at the base of the claws themselves (on trunk segment 1). In contrast, this species has them shifted more than 10 segments posteriorly. This heterotopic change and its consequence – of the venom having to travel through a much longer duct than usual to reach its target – is a particularly weird form of developmental repatterning. It's very hard to imagine what the selective advantage underlying it could be, and yet it doesn't seem like a type of repatterning that would be selectively neutral.

The land-snail *Cepaea nemoralis* can be used to exemplify both heterotypy and heterometry at the organismic level. Some snails in natural populations have yellow shells, some pink, and some brown – as a result of genetic differences between them. This is an example of heterotypy – different pigments. Also, superimposed upon the base colour of the shells is a variable number of black bands – from 0 to 5 of them. This is an example of heterometry in the form of an altered number. (Note that, in different contexts, heterometry may be more appropriately described in terms of quantity, rate, or concentration.)

Note that the axolotl and centipede cases exemplify *evolutionary* repatterning of development that happened in the birth of particular species, whereas the snail cases exemplify *genetic* repatterning of development resulting in phenotypes that coexist in the same species (and indeed in the same population). It seems likely that the former type of repatterning originates from the latter – in other words genetically based repatterning of development occurring within a species sometimes leads to speciation, though of course often it does not.

Developmental Constraint, Bias, or Channelling

The types of developmental repatterning discussed above might be described as theory-neutral. They are simply descriptions of kinds of alteration that evolution can produce from the starting point of a particular ancestral developmental trajectory. Things get more interesting when we begin to ask whether some types of developmental repatterning are somehow inherently more probable than others because of the (often unknown) dynamics of the developmental processes concerned. Recall that we first encountered this question in Chapter 1 in the context of leaf arrangements in plants. This issue of varying probability of generating variants is not just a question of, for example, whether it is easier to evolve by heterochrony than heterotopy, but also, for example, whether it is easier to evolve by some kinds of heterochrony than others. Indeed, since the four types of change can be intertwined, for example with heterotypy at the level of the gene causing heterochrony at the level of the organism (just one of many possibilities), we might want to reformulate our question in a more general form.

At its most general, the key question is this: from a starting point of a particular ancestral developmental trajectory, is the probability of generating any of an array of possible variant descendant trajectories (a) equal or (b) unequal? If the former, then the particular kind of developmental repatterning that occurs in a lineage is a *response* to a *driving mechanism* (probably natural selection); if the latter, then repatterning may be *part* of the driving mechanism (probably along with selection), rather than merely a response. This is an important distinction.

A big problem in getting to grips with this issue is that it is often hard to define the 'array' of possible variants. However, centipede segment numbers provide an interesting example where this is less hard than usual. In the geophilomorph (soil-dwelling) group of over 1000 species of centipedes, the number of trunk (= leg-bearing) segments is always odd. It can be as low as 27 or as high as 191. Almost every odd number within this wide range is found in at least one species. However, no even number characterizes *any* species. So, if the optimum number of trunk segments (and hence pairs of legs) for a particular species in a particular environment is (say) 42 from a fitness perspective, then the closest a species of geophilomorph could get to this number would be 41 or 43. Perhaps the best way to think of this situation is that natural selection can push the distribution of segment numbers into a general area (low forties) but the way in which segments develop 'constrains' it to odd numbers only within this area – we'll examine how this constraint operates in Chapter 5.

However, note that constraint is a rather negative term. We could equally well say, on a more positive note, that the way in which segments develop *drives* the number into the odd character states in all lineages. But maybe the best form of terminology would be one in which neither negativity nor positivity is implied. After all, the key question, as stated at the start of this section, is about the *relative probabilities* of generating some variants compared with others – across the whole array of conceivable ones. An increase in the relative probability of one variant implies a decrease in the relative probabilities of others.

Different authors have used different terms to indicate such unequal relative probabilities. Stephen Jay Gould was one of the many who used *constraint*. He argued that it could cover positive as well as negative aspects, pointing to the (archaic) usage 'I feel constrained to speak' (meaning 'I feel compelled to speak'). Other authors, including myself, have used *bias* – the developmental process may be such that there is a bias in favour of certain variants and against others. And other authors again, including the Swiss biologist Chris Klingenberg, in his 2010 review article on the evolution and development of shape, have talked in terms of developmental *channelling*.

Call it what you will – constraint, bias, or channelling – this is something that is very definitely *not* theory-neutral. I mentioned in Chapter 1 a controversial paper written by Stephen Jay Gould and Richard Lewontin in 1979; now we will discuss it in more detail. The main point these authors made was that the directions in which evolution alters development are not entirely due to natural selection, but rather owe much to what they called constraints (both developmental and other). Here is how they put it (my italics):

> organisms must be analysed as integrated wholes, with *Baupläne* [= body plans] so constrained by phyletic heritage, pathways of development and general architecture that *the constraints themselves become more interesting and more important in delimiting pathways of change* than the selective force that may mediate change when it occurs.

This was seen by many biologists as an attempt to dethrone Darwin. While the paper certainly wasn't an attack on Darwinism, it was an attack on the form of *neo*-Darwinism that sees all evolutionary change as being caused by the external environment – I call this pan-externalism, to distinguish it from more reasonable forms of neo-Darwinism, of which there are many. But is the claim made by Gould and Lewontin right or wrong, or is it merely a matter of opinion and slippery terminology? The best way to examine this is through case studies, and we will take that approach in the next chapter. However, there is one other aspect of the debate that should be noted right now – the difference between absolute and relative constraint.

An absolute constraint takes the form of apparently 'prohibited' phenotypes (or developmental trajectories, in the time-extended view). Even numbers of centipede trunk segments provide an example. Six-legged vertebrates provide another: zero is possible (e.g. snakes), two is possible (e.g. humans and bats), four is common, but six is non-existent – despite being fine from an engineering point of view, as insects prove. We have to be careful here: claiming evolutionary prohibition on the basis of past evolution is dangerous, given that we don't know what evolution will produce in the future. However, with that caveat in mind, an absolute constraint is clearly different from a relative one. When constraint is relative, the probabilities of generating two variant descendant developmental trajectories from a particular starting point are neither 50/50 (lack of constraint) nor 100/0 (absolute constraint) but rather somewhere in between (say 75/25). Relative constraint is almost certainly much commoner than its absolute counterpart. It can sometimes be overcome by selection and sometimes not, as we will see in the next chapter, thus making the nature of the interplay between the two (constraint and selection) a topic of considerable interest.

Modularity and Evolvability

The idea of 'evolvability' has been proposed by various authors, including, in 1998, the American biologists Mark Kirschner and John Gerhart. The evolvability of a developmental trajectory can be defined by the capacity of the species it characterizes to generate variants of it – ones that are both (a) heritable and (b) at least fit enough to be able to produce variant adult offspring in the wild. Although other definitions are possible, this one works reasonably well. Evolvability can thus be thought of as the opposite of constraint: the more constrained a system is, the less evolvable it is, and vice versa.

It has been proposed that there is a link between the evolvability of a developmental system and its modularity. A developmental module is a part of an overall developmental system that is quasi-independent of the rest of it. The emphasis here is very much on the *quasi*, because it is in the nature of developmental systems that everything is connected to everything else.

Nevertheless, with that caveat, the concept of modularity is useful. Examples of embryonic modules are limb buds and the primordia of various organs, for example eyes.

The basic idea is that, although everything is connected up in a developmental system, the overall pattern is such that there are far more causal interconnections within a module than between modules. This reduces the probability that changing a module will have disruptive consequences elsewhere in the developing organism. Across the vast majority of the animal kingdom, many modules occur as pairs, because of bilateral symmetry. In general, what is evolvable is not a module individually, but a pair of modules. For example, the tiny forelegs of a tyrannosaur are a result of the repatterning (heterometry: downsizing) of their development quasi-independently of the size of the body, including the hindlegs. But they are not independent of *each other*. However, that said, left–right asymmetry in the development of limb modules *can* occur – it's just less common. An example is the asymmetry of the claws in male fiddler crabs.

In the fish known as the Mexican tetra *Astyanax mexicanus*, populations living in caves have evolved towards being blind. Like arm development in tyrannosaurs, this is a case of downsizing of the relevant developmental modules. Interestingly, some populations of this species have reduced eyes and sight, while others have no eyes and are completely blind. Such fish typically have enhanced chemoreceptors (corresponding to enhanced smell/taste) on their heads. Clearly, chemoreceptors are more useful than light receptors when you live in an environment in which there is no light.

The evolutionary loss of a structure is often hard to explain. The fact that it is no longer useful in a new environment does not necessarily mean that selection will act against it. The structure might become vestigial and eventually disappear through genetic drift. Alternatively, selection might act against it because any structure has associated production costs, and if energy can be diverted away from it and used in the production of a more useful structure this combination may be selectively favoured. Perhaps this is what has happened in the case of the blind cave-fish. Whether this is the correct hypothesis or not, the fact that eye primordia are quasi-independent developmental modules has clearly been a factor in their evolvability, in the same way that quasi-independence has been important in the evolution of many sorts of limbs.

The Evo-Devo Hourglass

I mentioned in Chapter 2 that there is a modern version of von Baer's idea of developmental divergence, namely the developmental hourglass (or egg-timer). This provides an example of apparent lack of evolvability and therefore acts as a counterpoint to the above examples of modularity. It thus helps to define the other end of the *spectrum* of evolvability – and it most certainly is a spectrum rather than a dichotomy.

As we saw in Chapter 2, von Baer claimed that a comparison between the developmental trajectories of two animals belonging to the same large group – say two species of vertebrate – typically revealed a pattern of early similarity giving way to later differences. Our current view of the pattern revealed by such a comparison is more nuanced: the earliest stages of the life cycle often have some notable differences; later stages (but often not much later) are more similar, with the stage of greatest similarity being called the phylotypic stage; then stages that are later again show increasingly major differences. This is the case, for example, in a comparison between the development of a bird and a mammal. Mammalian and avian eggs are of course *very* different from each other. The early mammalian blastula stage (the blastocyst) is quasi-spherical, while the avian blastula stage (the blastodisc) is flattish, lying on top of a mass of yolk. Later on, the embryos converge to the form with gill clefts that is called the pharyngula (the neck of the hourglass, and hence the phylotypic stage of vertebrates), before then diverging in the way that von Baer showed, with the development of hair in one case, feathers in the other, and so on (Figure 4.1). This pattern was referred to as an 'egg-timer' in 1994 by the Swiss-French biologist Denis Duboule; later authors have used either egg-timer or hourglass.

Mammals and birds are direct developers; there are no cases of genuine larvae across any of the 15,000 or so species belonging to these two groups. The variation at the start of the developmental hourglass can be even more pronounced in the case of indirectly developing animals. In the previous chapter, I mentioned that sea-urchins of the genus *Strongylocentrotus* have been the basis for a case study of the evolution of feeding and non-feeding larvae. The former have 'arms' to help them acquire planktonic food, while the latter lack such arms and are fed from an internal yolk supply. Species of this genus can have very different larvae yet very similar adults. This kind of pattern can also be seen in some comparisons between different species of

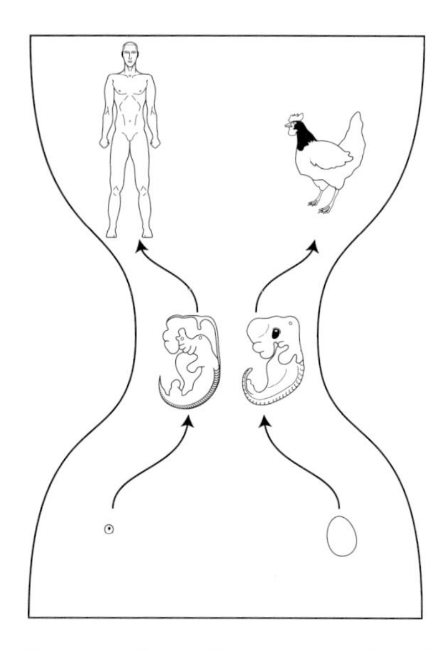

Figure 4.1 The evo-devo egg-timer or hourglass. In comparisons of the development of animals representing different groups, for example mammals versus birds, we often find a pattern of early differences giving way to later similarity and then to differences again. The point of constriction in the hourglass can be referred to as the phylotypic stage, which in vertebrates is the pharyngula. Note that this stage is much closer in time to the zygote than the adult; so, unlike a real hourglass, this conceptual one is very asymmetric.

lepidopterans. For example, the caterpillars of two species of the butterfly genus *Colias* – *C. hyale* and *C. alfacariensis* – are more readily distinguished from each other than are the adult butterflies.

Von Baer's idea of progressive embryonic divergence wasn't perfect; neither is the idea of an hourglass. When the hourglass pattern is found between direct developers, as in the bird–mammal comparison, it is usually very asymmetrical, as noted in the caption to Figure 4.1. When this pattern is found between indirect developers, as in the cases of sea-urchins and butterflies, we could argue that in fact such life cycles are really two linked developmental systems: embryo-to-larva and something-to-adult, and that each of the two systems, when considered on its own, really shows a von Baerian pattern. The 'something' in butterflies is the system of imaginal discs that we saw in Chapter 3; in sea-urchins it is a little piece of tissue within the larva called the rudiment, which is the developmental source of the adult. In some comparisons, for example between the developmental trajectories of two different species of mammals, it is perhaps the case that von Baer's progressive divergence is a *better* fit than an hourglass. This reinforces the point made several times already: the evolution of development is such that laws (such as von Baer's) are inapplicable. Instead of a fruitless search for universals, we should search for interesting repeated patterns, and the variations that are superimposed upon them.

The most evolutionarily conserved stage in the development of vertebrates is, as we have seen, the pharyngula. This particular phylotypic stage is not found in any animal phylum outside the chordates, but other phyla or sub-phyla often have an equivalent stage. In insects and other arthropods it is the germ-band stage. In echinoderms it is the rudiment stage. However, whether *all* animal phyla can reasonably be said to have a phylotypic stage is not yet clear. Indeed, the majority of the 35 or so animal phyla have been little-studied compared to chordates and arthropods; and some of them do not have a great enough number of extant species for the necessary interspecific comparisons to be made. For example, in the case of the phylum Placozoa ('flat animals'), there is so far just a handful of named species.

The defining feature of a large group of animals such as a class or a phylum may be a particular phylotypic stage in development, or a particular early embryonic structure, such as the chordate notochord, or a particular feature of subsequent developmental stages and adults, such as the vertebrate endoskeleton and the arthropod exoskeleton. Structures such as skeletons that characterize all (or virtually all) species in the large group concerned are referred to as body-plan features. These, like the phylotypic stages that precede them in development, seem to be very constrained in evolution – not in the sense that they can't be quantitatively modified (they clearly can be) but in the sense that they can't be dispensed with. The body-plan concept is a major focus of attention in evo-devo, and we now need to consider it further.

Body Plans and Evolutionary Novelties

'Body plans are easy to exemplify but difficult to define.' I used those words in an earlier (1997) book that was focused on the body-plan concept, and I think they remain true, despite an intervening 20+ years of evo-devo research. However, let's make the effort to come up with some sort of rough working definition.

The four essences of the concept are: (a) an overall body structure or layout in terms of general architecture; (b) an association with what can be called a 'higher taxon' in the sense of a level or rank of taxon that is at the high end of the taxonomic hierarchy, such as a phylum or sub-phylum; (c) a lack or scarcity of exceptions to possession of the structure concerned within the relevant taxon; and (d) an earliness of its origins in the developmental trajectory of the animals concerned. Combining these four essences, we can arrive at the following working definition: a body plan is the general pattern of organization of the body of animals belonging to a high-level taxon, which is shared among all or almost all of them, and has its origins in early development.

This definition works for the two examples given at the end of the previous section: the vertebrate and arthropod skeletons. Of course, when a word used in a definition is itself undefined (a commonly encountered problem) then the definition is correspondingly problematic. For example, how early is 'early'? Skeletal rudiments are not to be found in the vertebrate blastula, for example. They do not make their appearance – in the form of cartilage – until much later. However, cartilage comes from the mesoderm germ layer, which makes its appearance at the gastrula stage, which immediately follows the blastula. Not long after the gastrula stage, there is partitioning of the mesoderm into sub-populations of cells, including a population called the sclerotome, in which stem cells will progressively divide, with some of their daughter cells becoming chondrocytes – the cells that make cartilage.

Does the definition also work for other examples of body plans? Let's consider the phylum Echinodermata – comprising the sea-stars, brittle stars, sea-urchins, sea-cucumbers and sea-lilies. These are characterized by radial symmetry – usually pentaradial (fivefold). This is clearly an overall body-layout feature. There are no exceptions to radial symmetry in the present-day echinoderm fauna, though the fossil record shows that the first echinoderms did not have pentaradial symmetry – hardly surprising, given that they shared a bilaterally symmetrical ancestor with chordates. As we have seen, the echinoderm adult is constructed from the piece of tissue in the larva called the rudiment. It is in this that the pentaradial symmetry has its developmental origins – the larva itself has bilateral symmetry.

The body plan of the Mollusca involves a muscular foot, a visceral mass, and a sheet of thick skin called the mantle. Cells in the mantle secrete the shell, present as a single spiral form in snails, two hinged pieces in the bivalves such as cockles and mussels, a series of eight plates in the chitons, and in various other forms, including reduced/internal as in the squid and cuttlefish. Some molluscs have lost their shells, evolutionarily speaking (e.g. slugs, sea-slugs, octopuses), but none has lost the mantle.

If we use 'body plan' specifically for the phylum and sub-phylum levels of the taxonomic hierarchy, what term, if any, do we use for the next levels down, namely classes and orders? There are at least three possibilities. First, since the German word for body plan, *Bauplan*, is sometimes used in the (English) literature, we could use *Unterbauplan* for the next level down. Although some authors have done this, I feel that it suggests too rigid a pattern, especially given that the various ranks of taxon do not have precise definitions. Second, we can use the term 'evolutionary novelty' when a class or order is characterized by some prominent feature that arose in the stem lineage of the group concerned. One of the best examples is the shell of turtles and tortoises (order Testudines). However, not all orders and classes have such a clearly novel structure. In such cases we can take the third possible approach and use the cladistic term 'derived characters' for features that are associated with the group – for example the pattern of two upper and two lower (continuously growing) incisor teeth that characterizes the order Rodentia.

One of the key questions in evo-devo is whether the origins of evolutionary novelties and body plans involve processes that are different from those that are involved in routine evolutionary change, such as the proliferation of different species of Darwin's finches. If the answer to this question is 'yes', then this would suggest either that (a) there is something unique about phyla, classes and orders rendering their evolution distinct from that of lower-level taxa, which seems unlikely, or that (b) every level in the taxonomic hierarchy is different from every other one in terms of the evolutionary processes that generate them, which also seems unlikely. On the other hand, if the answer is 'no', then our conclusion must be that the sorts of evolutionary change that lead to different beak sizes in birds are essentially the same as those involved in the origin of beaks in the first place, and those involved in the origin of even more major features of animal body layout, such as bilateral symmetry, a skeleton, and anteroposterior segmentation. This seems equally unlikely. So we have a problem either way.

The solution to this problem may lie in statistical but not absolute differences, as follows. It seems likely that the evolutionary origin of a new body plan typically involves earlier developmental changes than those that characterize a typical speciation event that leads to very similar sister-species, for example the eastern and western gorillas. If this is true, then it may mean that there is also a statistical difference in the kinds of developmental genes involved (see next section). Perhaps at the population level there is also a statistical difference in the relative importance of natural selection and other factors, including developmental bias.

Genes for Building Bodies

We have already seen that genes can be grouped into three classes: developmental, cell-type specific, and housekeeping. In their 2005 book *From DNA to Diversity*, the American evolutionary developmental biologist Sean B. Carroll and his co-authors extend the house metaphor beyond one category – to include all three of them. Thus in addition to the housekeeping genes that are constantly ticking over all around the house, Carroll *et al.* describe cell-type specific genes as those that carry out specialized tasks in individual rooms (cell types), and developmental genes as those that built the house in the first place. The last group – the most important for us – could be called 'building genes' (a term that is not often used), 'developmental genes' (which I have used a lot up to now), or 'toolkit genes' (which is what Carroll *et al.* and many other evolutionary developmental biologists use). Whether in fact 'developmental' and 'toolkit' genes are two names for exactly the same thing we'll come back to later. For now, the important thing is to have a broad classification of genes, to give a general background to discussion of genetic aspects of evo-devo – see Figure 4.2.

As can be seen in the figure, the key category of developmental – or toolkit – genes can be subdivided into genes that make transcription factors (proteins that switch other genes on) and genes that make the components of cell–cell

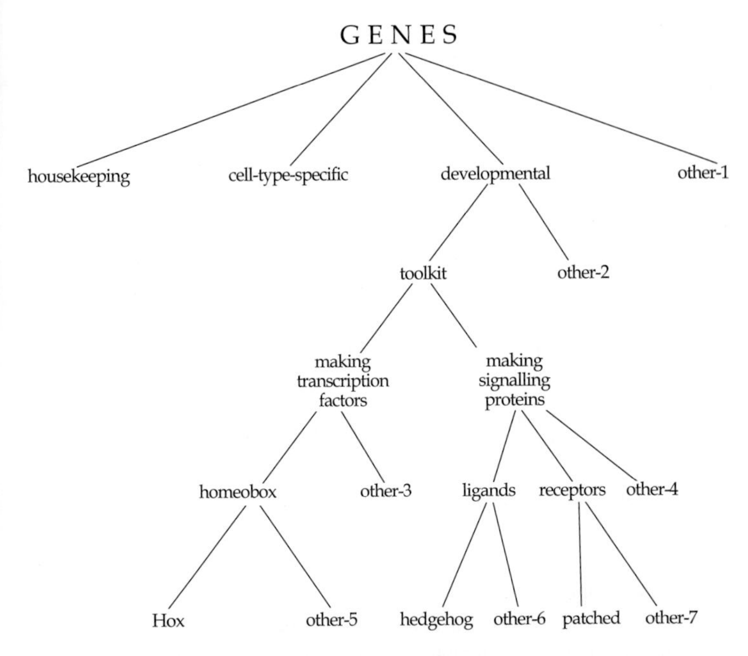

Figure 4.2 Hierarchical classification of genes, with a particular focus on those that are involved in development. Important types of developmental gene that we have already considered, for example homeobox genes, can be found in the diagram. When we later encounter further important developmental genes, I will relate them back to this classification. Note that for each split, I have included an 'other' category. This is in some cases because the named categories are not exhaustive, and in some cases because they may not be so. These various 'others' are numbered so that I can refer to them later.

signalling pathways (involving proteins that are secreted out of one cell and have developmental consequences when they are received by another). Although signalling pathways and transcription factors are in constant interplay, having a mental distinction between the two is useful. Each of these subcategories can then be split further. Homeobox genes, with which both evo-devo and this book started, are a major category of genes that make transcription factors; but there are others too, whose protein products bind to DNA but do so via a structural motif that is different from the homeodomain.

Signalling pathway genes can be broken down into those that make the secreted protein (which is generally called the ligand), those that make transmembrane receptors for the ligand, and those that make other components of the pathway.

The homeobox genes can be further split into Hox genes and others. When first introducing homeobox genes in the introductory chapter, I avoided mentioning the Hox genes because there were already enough new terms to get used to. But now we need to examine this important category of developmental genes. Although 'Hox' sounds like simply an abbreviation of 'homeobox', it is not. This annoying fact is an outcome of the rather anarchical, quasi-historical way in which new biological terms originate. The broader category, homeobox genes, can be defined very simply as all genes that contain a homeobox sequence (about 180 base-pairs long, as we saw earlier). The narrower Hox category is harder to define, but essentially a Hox gene is one that not only has a homeobox but is also (a) a gene controlling development along the anteroposterior axis of a bilaterian animal, (b) potentially subject to homeotic mutation, and (c) usually part of a gene complex or cluster occupying a short stretch of a chromosome that contains other Hox genes – the number depending on the animal concerned (flies have eight of them, while humans have 39, organized into four separate clusters). In most animals, Hox gene clusters show a fascinating phenomenon called colinearity in which the order of the genes along the chromosome is the same as (or very similar to) the anteroposterior pattern of their expression in the body. The reason for colinearity is still not completely understood.

The genes making the ligands and receptors of cell signalling pathways can also be split. Regarding ligands, we have already come across two members of the Hedgehog family of genes/proteins – the *hedgehog* gene of *Drosophila* and one of the human Hedgehog-family genes – *Sonic hedgehog*. Because of this, I have singled out this family in Figure 4.2 from 'other-6'. The latter category contains genes for many other ligand proteins, including those of the Nodal and Notch pathways, which we will meet later (in Chapters 6 and 8 respectively). Also, genes for receptors can be split in a way that parallels the split in ligand genes. So, corresponding to the split between Hedgehog and 'other-6', we have a split between genes making the receptor protein Patched and 'other-7', because Patched is a transmembrane receptor for Hedgehog. In general, different ligands have different receptors.

In later chapters we will meet various toolkit genes that have not been discussed so far in the book. When we first encounter each of these I will state where it falls within the categorization of Figure 4.2, so that it has some context and is not just an isolated name hanging in the proverbial ether. Although the figure provides much-needed context, it should not be seen as either perfect or comprehensive. All classifications – of genes or any other biological entities – are simplifications of reality. Some genes may belong to more than one category. Some of my 'other' categories hide a wealth of detail. The idea that a particular ligand has a particular receptor is a considerable simplification. In reality, two or more receptors may work together. And so on and so forth.

Finally, an important distinction that is not captured in Figure 4.2 is the distinction between coding and non-coding DNA. There are many types of the latter. Some of it is *in addition to* the genes. So, for example, when we say that the human genome – the totality of human genetic material – has about 20,000 protein-coding genes, that statement takes no account of parts of the genome that are non-genic. The relative sizes of the genic and non-genic parts of the genome vary among different kinds of organism. The reasons for this enormous variation are not yet clear, though we do know that there are many different types of non-genic DNA.

Even more important than the division of the genome in this way is the division of an individual gene. A gene does not contain just the DNA sequence that gets transcribed into its RNA message – the coding region of the gene – it also contains non-coding sequences. These include regulatory sequences, some of which are binding sites for those transcription factors made by other genes that turn on the gene concerned. Evolution involves changes in both regulatory and coding sequences, and the relative importance of the two is of considerable interest. It is possible that most evolution of animal form is driven by changes in the regulatory sequences of developmental genes rather than in their coding sequences; Sean Carroll and others have argued in favour of this hypothesis.

5 The Evolution of Variations on a Theme

Levels of Evolutionary Change

Rapid evolution can be observed happening in nature when selection is unusually strong. We are all familiar, these days, with the evolution of antibiotic resistance in bacteria and the evolution of pesticide resistance in insects. Less familiar, but also very rapid, is the evolution of resistance to heavy metals in populations of plants that have adapted to growing on the spoil-heaps surrounding zinc and lead mines. These cases of unusually strong selection and consequently rapid evolution are all associated with human modification of the environment. The classic case study of evolution happening – industrial melanism in moths – also fits into this category.

It is generally agreed that these cases of rapid evolution, where the changes involved took anything from a few years to a few centuries, differ from their slower counterparts where selection is weaker – for example slight evolutionary changes in body form in *Homo sapiens* that have taken place over the last 100,000 years – only in terms of speed. The nature of the process is the same. Whether evolutionary change within a species happens quickly, slowly, or at a moderate pace, the essence of the situation is natural selection acting on variation in the populations concerned, and modifying that variation in the direction of enhanced fitness.

Can this conclusion, that evolutionary processes are essentially the same as each other except for their rate of occurrence, be scaled up from changes within a single species to the divergence of different species and different higher taxa? To say that there is a lot in this question would be an understatement *par excellence*. At this point we need to revisit the terms microevolution and macroevolution, which we first encountered in Chapter 2 in relation to the work of Richard Goldschmidt. We also need to examine the related, but less-often used term megaevolution, which was introduced by the American palaeontologist George Simpson in his classic 1944 text *Tempo and Mode in Evolution*.

The cut-off between microevolution and macroevolution is speciation – or 'the origin of species', as Darwin straightforwardly called it. For those who, like Goldschmidt, use only the two prefixes micro- and macro-, everything that goes beyond the confines of a single species is macroevolution. However, for those who, like Simpson, use all three prefixes, there is another cut-off, but it's a harder one to define. Megaevolution is typically thought of as the divergence of higher taxa such as classes and phyla. Thus the 2016 book *The Origin of Higher Taxa*, by the British palaeontologist Tom Kemp, is a study of megaevolution, even though he does not use this term.

Perhaps a better way to define megaevolution is in terms of body plans. Evolution 'within a body plan' is microevolution (if it is also within a species) or macroevolution (if it isn't); in contrast, the evolutionary changes that produce a new body plan constitute megaevolution. Here is what Simpson said about the differences between the three levels of evolutionary change:

> The palaeontologist has more reason to believe in a qualitative distinction between macro-evolution and mega-evolution than in one between micro-evolution and macro-evolution.

Nearly three decades after Simpson wrote those words, palaeontologists became embroiled in a debate over the idea of punctuated equilibrium, introduced by the American palaeontologists Niles Eldredge and Stephen Jay Gould. This idea – that almost all morphological (and hence developmental) change occurs during speciation events – focused attention on the distinction between microevolution and macroevolution. There are still different ideas on whether or not punctuated equilibrium is the norm, and if so, why. My own view is that a pattern of long periods of morphological stasis punctuated by lineage splittings that appear to be relatively rapid and involve notable morphological change is indeed found, though by no means universally. However, we need to pay attention to the context in which punctuated equilibrium is studied. This is normally the examination of a temporal series of fossils from the same place. I suspect that the punctuated pattern appears because what is being observed is not an evolutionary change at that place (the odds are against it) but rather an after-effect of a change occurring elsewhere. If this is true, then the 'punctuation' at the place observed is an ecological event (invasion) rather than an evolutionary one (speciation), which means that the speed of the evolutionary event remains unknown, and may have been fast, slow, or somewhere in between.

So, my view is that the debate over punctuated equilibrium was the proverbial 'storm in a teacup', and that it distracted attention away from the more important issue of whether megaevolution is in some way different from macroevolution, which I focus on here. Because I share Simpson's view that there is more reason to believe in a distinction of this kind than at the lower level where the debate over punctuated evolution was focused, I have combined evolution at micro- and macro- levels into this chapter rather than devoting a separate chapter to each. Evolutionary processes described here can then be collectively compared with those described in the following three chapters (6 through 8), which deal with the evolutionary origins of novelties and body plans. These origins seem to involve processes that are developmentally different from what might be called 'ordinary macroevolution', such as occurs in the proliferation of species within a family (e.g. the cat family Felidae or the crow family Corvidae). That said, you should not anticipate either: (a) a claim that the causes of the evolutionary events discussed in this chapter and the causes of those discussed in the next three are non-overlapping sets; or (b) a claim that the causes are indistinguishable in all respects. The truth, I suspect, will turn out to be more nuanced (and more interesting?) than either of those extreme views.

Developmental Variation

The starting point for all evolution is the existence of variation – but what forms does it take? And, in relation to the big question about whether evolution works the same way regardless of level, does the same kind of variation feed into evolutionary changes regardless of whether these will lead to new body plans or never transcend the species? Another important question is whether variation is in some sense biased, with some variants being 'easier' to produce than others.

This last question – which we began to examine in the previous chapter – is a key one in current evo-devo; and it has a long history of being asked by evolutionists of various hues. It was lurking in the background when Darwin talked about 'correlation of growth' in *The Origin of Species*, and it was implicit in a point made by J. B. S. Haldane, one of the founders of the 'modern synthesis' of evolutionary theory in the mid-twentieth century. In his 1932 book *The Causes of Evolution*, Haldane said:

> To sum up, it would seem that natural selection is the main cause of evolutionary change in species as a whole. But the actual steps by which individuals come to differ from their parents are due to causes other than selection, and in consequence evolution can only follow certain paths. These paths are determined by factors which we can only very dimly conjecture. Only a thorough-going study of variation will lighten our darkness.

He was – and is – absolutely right.

Before looking at case studies on the issue of whether variation is in some sense biased, we should note a couple of important points about variation in general. First, what matters most to evo-devo studies is developmental variation. Variation in housekeeping genes and resultant phenotypic variation that concerns metabolic but not developmental pathways is of less concern. Second, developmental variation has a mixture of genetic and environmental causes. In any particular case, the heritability can be anywhere from zero (environmental causation) to 100% (genetic causation). Cases of zero heritability are unimportant, as natural selection can do nothing with them. All the others *are* important; heritability values for developmental variation in particular populations have a wide range, often between about 50% and 90%.

The division into 'genetic' and 'environmental' causes is a simplification, even when given as a continuum of percentage values rather than as a simple binary split. There are many forms of interaction between genetic and environmental factors. One approach to this important area is via what are called developmental reaction norms. The idea here is that, for an animal of any particular genetic constitution, the value that a character (e.g. adult body size) ends up with depends on the value an environmental variable (e.g. population density) takes during development. For example, in a population of an insect species in which all the individuals are genetically identical, for example due to clonal reproduction (as can occur in aphids) or inbreeding by an experimenter (as is common in fruit-flies), those adults that result from culturing at

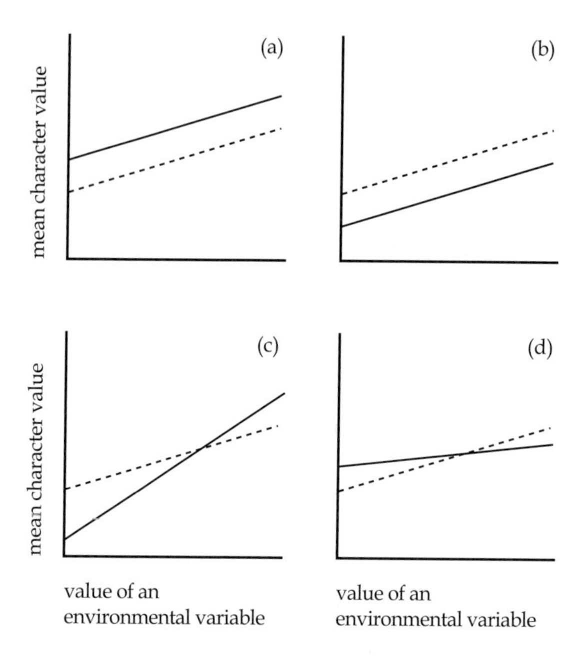

Figure 5.1 Four possible results of selection on patterns of phenotypic plasticity. In all cases, the starting point is the dashed line. Top (a and b): slope remains the same. Bottom (c and d): slope is altered – in one case (c) steepened, in the other (d) flattened. Note that more complex patterns of plasticity than linear ones are common.

low larval density end up with larger adult body size than those that result from culturing at high density.

In such a genetically homogeneous population, we are seeing *only* the environmental effect on body size. However, in a genetically variable population, selection can alter the pattern of plasticity (Figure 5.1). This shows that the way in which a character responds to the environment (i.e. the reaction norm) is itself genetically determined, at least in part. In Figure 5.1 the variation in the character is continuous, but this is not always the case in interactions of this kind. We saw an example of a discontinuous character in Chapter 2, with Waddington's selection on 'crossveinlessness' in fruit-flies. This discrete character was only produced after an environmental shock before selection; after selection it occurred spontaneously, because the frequency of genes giving that phenotype an enhanced probability of occurrence had been increased by selection.

The term 'character' is much in evidence in the above paragraph, and it will continue to be central to the discussion as we proceed, so a few words are necessary to clarify what this means. Basically, a character is some measurable feature of an organism. The example used above was adult body size, but it does not need to be a feature of the adult – rather, it can be a feature of any stage of the life cycle. For example, in a fruit-fly, we could refer to 'the length of the larva just before it pupates' as a character. We would expect this to be positively correlated, but not perfectly so, with the body length of the resultant adult fly. Also, as noted above, a character does not have to be something that varies in a continuous manner in a population. The same applies to variation between populations, and indeed between species. In cladistics it is common to talk of different 'character states' when referring to discrete variation between species. For example, the dorsal appendages on the third thoracic segment of an insect can be wings (in most insect groups) or the small flight-balancing organs called halteres (in flies) that we encountered in Chapter 2 in the context of the bithorax mutation. So, even without considering other possibilities (such as winglessness), we can see that this character, namely 'flight appendages on the third thoracic segment' has at least two discretely different character states in insects, namely 'wing' (W) and 'haltere' (H).

In cladistics, a question of much interest is which character state is ancestral and which is derived, and in this example we know the answer because of the pattern of distribution of the two states across the winged insects. W was the original character state in the large group called the Pterygota (winged insects), while H was derived from W in the origin of the flies (order Diptera). It is interesting to note, in this context, that there is another group of insects (order Strepsiptera) in which the forewings (from the second thoracic segment) have been turned into flight-balancing organs, rather than the hindwings (from the third). This structural arrangement has been much less successful than that of the flies, as judged by the number of species in today's fauna – fewer than 1000 species of Strepsiptera but more than 100,000 species of Diptera. However, whether this big difference in success is due to the wing pattern or to other factors (or both) is not clear.

Returning to developmental variation within a species, which is the starting point for all morphological evolutionary change, this is more often continuous than discrete. It typically takes the form of a normal distribution of character values, as in the case of adult human height – though here, as in many other characters and species, the normal distributions differ between the sexes. This kind of distribution is typically the result of variation at many genes, plus environmental effects (and genotype–environment interactions). The genes involved in a particular continuous character (or trait) are referred to as polygenes, simply because a lot of them work together to influence the character concerned. The individual effects of each polygene are typically small, and the identities of the genes often remain unknown. However, their chromosomal locations can be mapped.

The stretches of chromosomes that include one or more polygenes are referred to as quantitative trait loci, or QTLs. The number of these responsible for the variation in a particular character is often in the range from about 10 to about 50. However, they have unequal magnitudes of effect. This means that specifying the exact number is difficult. In a particular case, there might be only 10 QTLs that have an effect bigger than *X*, but 30 that have an effect bigger than *Y*, where *Y* is less than *X*. One aim of the discipline of quantitative genetics is to map as many as possible of the QTLs that contribute to the variation in a character such as the body length of a mouse, or, in a medical context, susceptibility to a particular disease. Usually, they are scattered all over the genome rather than being tightly clustered on a particular chromosome. (This is the first time we have encountered 'quantitative genetics'. This name refers to the branch of population genetics (*sensu lato*) that deals with continuously variable characters.)

It's important to realize that 'polygene' is a heterogeneous category in terms of what the genes do. Hence it does not fit neatly into the gene classification shown in Figure 4.2. Rather, different polygenes fit into different categories. Some of them are toolkit genes. This might at first seem odd, because we picture toolkit genes as those with major effects on development. However, it is important to distinguish a gene from a particular mutation of it. A gene that is subject to major-effect mutations – for example homeotic ones – may also be subject to minor-effect mutations.

An example of this is provided by the *Pitx-1* gene in sticklebacks. This is a homeobox gene, so it makes a transcription factor that switches other genes on. It has multiple roles, as is common in toolkit genes, and is found in a wide range of species, including humans. It is expressed in the pituitary gland (hence its name) and in limb development. Mutations in it can have major effects. But they can also have relatively minor ones. The stickleback species *Gasterosteus aculeatus* is found in a wide range of aquatic habitats. In some of them it has a well-developed pelvic skeleton, with pelvic spines projecting from it. In others, the pelvic skeleton is less well developed and the spines are reduced or missing. This variation seems to represent adaptation to different balances of selective forces in different habitats, including different levels and types of predation, with the spines being an anti-predator device. Crosses between individual fish from different populations have shown that one of the QTLs contributing to the variation maps to the site of the *Pitx-1* gene.

While quantitative genetics is concerned with analysing the QTLs contributing to developmental variation, the related subject of morphometrics is concerned with describing the variation at the phenotypic level. This can be done in simple ways such as using a ratio of one measurement to another one in order to describe a shape. For example, the breadth divided by the height of a snail's helico-spiral shell gives a measure of how 'squat' it is. Alternatively, a more sophisticated method can be used for more subtle shape measures, based on the relative arrangement of a set of identifiable points or 'landmarks'– e.g. on the complex and irregular structure of a bone. In both cases, the focus of interest is on how the shape measure concerned changes in both development and evolution, and, in the latter case, the nature of the selection that brings about the changes.

We now return to the issue of whether developmental variation is in some way biased (or constrained or channelled). Let's consider the variation in two characters in land vertebrates: 'length of adult forelimb' and 'length of adult hindlimb'. These tend to be positively correlated, both within and between species. Thus a human with legs that are longer than average will typically have longer arms too. Equally, a giraffe has both longer forelegs and longer hindlegs than the mammalian average; and a shrew has both pairs of legs shorter than average. Such correlations indicate that developmental variation is *structured* (or biased) rather than random. Note however that there are exceptions to these correlations, such as tyrannosaurs with their massive hindlegs and vestigial forelegs.

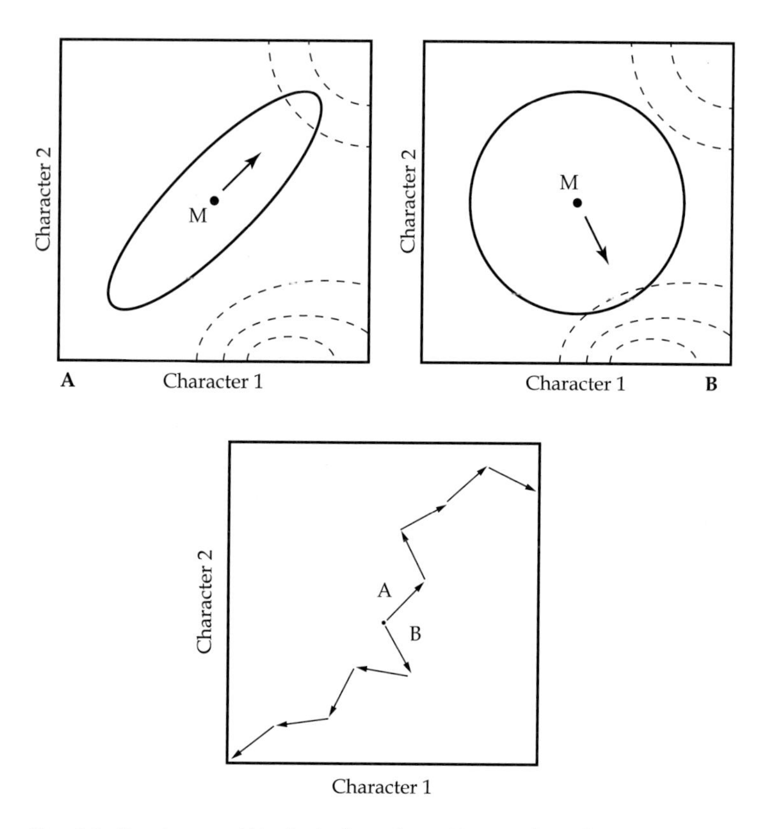

Figure 5.2 Developmental bias in the form of a positive correlation between two characters (top left, A) and lack of bias (zero correlation, top right, B). Also shown (bottom) are the long-term consequences that follow from whether a population evolves as in A or B in the short term. The point marked M is the population's combined mean value, the ellipse in A and the circle in B delimit the variation, and the dashed lines are fitness contours.

Figure 5.2 shows a case of biased variation (specifically a positive correlation) involving two characters (top left) and a case of unbiased variation (zero correlation, top right) for contrast. The dashed lines show fitness contours representing hills that can be called adaptive peaks – though in fact the actual summits (fitness maxima) are off the pictures; it is the adaptive *slopes* that are shown. The main part of each of the top two pictures is a flat area of equal fitness – variation here is selectively neutral. The arrows show the direction of microevolutionary change; it can be seen that the direction of change is different in the two cases even though the population's combined mean value (M) of the two characters concerned is initially the same and the fitness landscape is also the same. This contrast shows how a pattern of character covariation can alter the direction in which evolution goes. This is another aspect of the blindness of natural selection: here, the patterns of variation mean that selection cannot 'see' one of the adaptive slopes – a different one in the two cases.

If selection drags a population up 'adaptive slope 1', the broader-scale fitness landscape that becomes visible may be different from that which would have been visible if instead it had dragged the population up 'adaptive slope 2'. Thus while the initial difference between going up slope 1 versus slope 2 might have been small in phenotypic terms, that difference might lead to a whole series of subsequent differences, as shown in the bottom part of the figure (where A and B are the initial steps depicted in the top part). Thus the long-term consequences of developmental bias may be profound.

We will now look at some examples of developmental variation and evolution, to observe the intertwining roles of developmental bias, natural selection, and other factors. We will start by looking at laboratory populations of a single species (a butterfly), then we will discuss a single species (of centipede) in the wild. Then we will go beyond the confines of microevolution: we will consider a pair of sister-species (snails), and finally we will look at a large clade of 5000+ species (mammals). In all cases, the evolutionary changes can be described as variations on a theme, though in the case of the mammalian clade the idea of a common theme becomes a bit strained in some of the lineages concerned, notably those that have taken to the air (bats) or returned to the sea (e.g. whales).

Artificial Selection and Developmental Bias

Artificial selection for certain character values carried out by humans has been very informative about natural selection ever since *The Origin of Species*, in which Darwin arrived at the latter from the former. In his day, artificial selection was practised for the most part by animal and plant breeders. However, by the middle of the twentieth century, selection experiments were often conducted by biologists, and this tradition has continued into the present century. Here I will focus on an interesting series of experiments carried out in the Netherlands by the British biologist Paul Brakefield, the Portuguese biologist Patricia Beldade, and their colleagues, published in two linked papers in 2002 and 2008. This work involved a species of butterfly called the African brown *Bicyclus anynana*, belonging to the lepidopteran family Nymphalidae, which includes many familiar species (e.g. the tortoiseshells). Overall, there are more than 5000 species in this family, and interestingly the family is characterized by the use of only four legs – the front legs have been reduced to small non-walking structures, though the nature of the selection that caused this change is not clear.

Like all butterflies, adult African browns have two pairs of wings, and, as in many species of butterfly, for example the confamilial European peacock *Aglais io*, the wings have pigmentation 'eyespots'. In the African brown, these vary in number, size, and colour. This variation has both genetic and environmental components. There is seasonal variation: wet-season forms have much more prominent eyespots than dry-season forms, this being a response to the temperature prevailing during development – with the wet season being characterized by higher temperatures than the dry season. But there is also genetic variation among individuals, and this can be used by experimenters in the lab (and by natural selection in the wild).

The Brakefield–Beldade group carried out two series of selection experiments that are especially relevant to the issue of whether developmental bias affects the direction of evolution, as suggested in general terms in the previous section. It thus represents an important case study on this issue. Both series of experiments involved the two prominent eyespots on the forewing. The sizes of these are correlated – bigger eyespot A goes with bigger eyespot B; also, the pigmentation patterns of the two are correlated – blacker eyespot A tends to go with blacker eyespot B. So in both cases the pattern of covariation is elliptical (as in the top left panel of Figure 5.2) rather than circular.

The experimenters tried to select in four directions both with eyespot size and eyespot pigmentation: the two 'easy' directions favoured by the elliptical pattern of covariation (the long axes of the variation seen in

Figures 5.2 and 5.3), and the two 'hard' directions in which there was less available variation (short axes). However, the results of the parallel experiments on the two characters were different: while selection was able to break the developmental bias that shows up as a correlation in eyespot sizes, it was not able to break the equivalent bias/correlation in eyespot pigmentation (Figure 5.3). So, the general conclusion would seem to be that in some cases of developmental bias, the bias will affect the direction of evolutionary change, while in others it will not, because selection can override it. In other words, sometimes the direction of evolutionary change is determined by selection alone, while sometimes it is a result of the interplay between natural selection and the structure of the available variation.

Beldade, Brakefield, and others have gone on to identify the developmental genes that underlie the observed phenotypic variation. As might be expected, many different genes are involved. One of them is the gene called *distal-less*, which, as we will see later (Chapter 8), is involved in the development of animal limbs across many taxa. Not only does this gene also have an important role in eyespot formation in *Bicyclus anynana*, but this role (it is sometimes called an eyespot activator) appears to be conserved across the family Nymphalidae, and perhaps beyond. This is an example of a toolkit gene that has more than one function in animal development; and it has become clear than multi-functionality of toolkit genes is the rule rather than the exception. Acquisition of novel functions may go hand-in-hand with the origins of novelties and body plans, as we will see in later chapters.

Phenotypic Plasticity and Developmental Bias

In Chapter 4, we noted the existence of a developmental bias in segment number in a clade of centipedes – the Geophilomorpha – that includes more than 1000 species. Here, we take a closer look at one of these species, namely *Strigamia maritima*. As its name suggests, this is an exclusively coastal species, which is rather unusual for a centipede. The geographic range of *S. maritima* is from Scandinavia to Iberia. It is found on Atlantic coasts as well as those of the Baltic Sea, the North Sea, and the Irish Sea, but its range does not seem to extend into the Mediterranean. Its closest relatives are other species of the genus *Strigamia* – these are 'normal' inland species, only occasionally found at the coast. It seems that the few coastal species within this group of 1000+

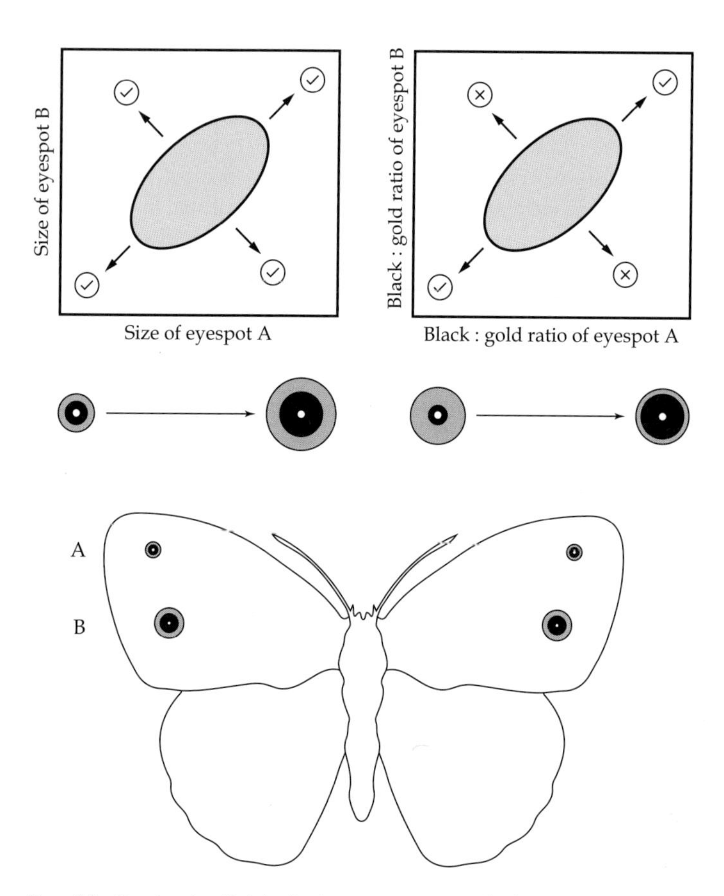

Figure 5.3 Results of artificial selection experiments on the butterfly *Bicyclus anynana*, which attempted to break developmental biases that show up as patterns of covariation involving two wing eyespots, A and B. Left: covariation in the sizes of A and B can be broken by selection (top left and bottom right arrows with ticks). Right: covariation in the extent of black versus gold pigmentation in A and B cannot be broken (top left and bottom right arrows with crosses). Gold pigment is here represented by grey. Below: dorsal view of an adult of this species, showing the positions of the two eyespots on the forewings. As indicated, the posterior eyespots are typically bigger than the anterior ones in natural populations.

have all evolved independently from inland ancestors – an interesting case of parallel/convergent evolution.

We noted in Chapter 4 that the range of the number of leg-bearing segments (LBS) in this group of centipedes as a whole is very broad – from 27 to 191. Within this overall range, an individual species usually has a narrow range of contiguous odd numbers. The cause of the developmental bias that produces the odd-numbers-only pattern appears to be the developmental production of segments via a stage involving double-segment units that later get split in two. This involves the intercalation of secondary stripes of expression of the centipede homologue of the *Drosophila* toolkit gene called *caudal*, as demonstrated by Ariel Chipman and colleagues in 2004. Development of segments from double-segment starting units should produce an even-only rather than odd-only pattern, the opposite to what is found. However, the venom claws of centipedes are evolutionarily modified legs, so the segment that bears them (the one that is between the head and the first LBS) should be included in counts of LBS, although traditionally it has not been.

In *S. maritima*, the number of LBS can take any odd value between about 43 and 53. The variation between these extremes can be broken down into several components. In any one geographic population there is sexual dimorphism, with females typically having a modal value that is the male modal value plus two. In addition, there is within-sex variation, often straddling three numbers. For example, a population might have males with 45, 47, or 49 LBS, females with 47, 49, or 51. The sexual dimorphism is genetically based (though the details of this are unknown); the within-sex variation is only partially heritable. Over the whole species range there is also between-population variation, which can be divided into latitudinal (individuals in more southern populations have more segments on average) and 'other' (because latitude does not explain all of the inter-population variation).

The sister-species of *S. maritima* – in other words its closest relative – is not known with certainty, but a plausible hypothesis is that it is an inland species of the genus *Strigamia* with a rather similar number of LBS. If this hypothesis is correct, then the speciation event that led to the origin of *S. maritima* may well not have involved a noteworthy change in modal segment number. Although modal number has changed in *S. maritima* as it has spread from an unknown point of origin across most western European coasts, these changes have not necessarily been selectively driven. We know that the restriction to odd numbers is due to developmental bias. And the latitudinal effect is potentially explicable in terms of the direct effect of temperature on LBS number (a form of plasticity), which, as was noted in Chapter 1, has been demonstrated by allowing eggs to develop into hatchlings (which have the full adult LBS number) at different temperatures. So we have evidence for both bias and plasticity; but as yet there is no evidence of selection on this character in this species. However, as usual, absence of evidence is not evidence of absence.

It seems likely that natural selection has been involved in the evolution of segment number in the Geophilomorpha as a whole. Within any one species, the variation may be selectively neutral for much of the time – that is, its genetic component is not acted on by selection. This may be the case in *S. maritima*, for example, though we can't be sure. But while the difference between 47 and 49 LBS *may* be invisible to selection, the difference between 27 and 191 (the extremes for the group) would certainly not be, for various reasons including the link between body length and number of legs: the body of a species that typically has 27 or 29 pairs of legs is simply not long enough to accommodate 191 pairs.

Natural Selection and Genetic Drift

The choice of case studies to examine greatly influences what we see. For those of us who are practitioners of evo-devo, there is a tendency to choose case studies where there is something of interest from our perspective, such as a particular form of developmental bias. But there is a risk that in doing so we end up overstating the role of such evo-devo processes in evolution in general. So it is prudent to include some case studies that are well known from an evolutionary perspective, but not from an evo-devo one. Here we will look at one such study – that of the evolution of the sister-species of European land-snail, *Cepaea nemoralis* and *C. hortensis*. These species have been much studied from an ecological genetics perspective, in relation to the visible polymorphisms manifested in the pigmentation of their shells. These polymorphisms are known to be heritable and to have a relatively simple genetic basis, involving a small number of genes.

We looked at the range of phenotypes in terms of shell colour and banding in *C. nemoralis* in the previous chapter. Those of *C. hortensis* are broadly similar both in varying shell colour (yellow, pink, brown) and varying number of dark bands (up to five). This suggests that the polymorphisms pre-date the speciation event that led to the divergence of *C. nemoralis* and *C. hortensis*. Hence it is likely that the polymorphism is actively maintained by balancing selection. One hypothesis for the nature of this selection is that bird predators, such as thrushes, form a 'search image' for the commonest phenotype, and as a result end up eating that form disproportionately. In such a situation, rare phenotypes are at an advantage; this is referred to as frequency-dependent selection. Interestingly, there is also evidence that avian predators can cause directional selection, with the result that the most cryptic phenotypes are selectively favoured – brown unbanded shells in beech-woods and yellow banded ones in grassy habitats. Thus the situation is complex in that the same selective agent – a predator – is thought to be capable of acting in ways that would tend to (a) maintain and (b) eliminate the polymorphism.

Predation is not the whole story here. Different shell phenotypes take up heat at different rates and thus produce different internal temperatures; so climatic selection is also involved, with paler shells being commoner in southern (Mediterranean) populations, less common in northern (Scandinavian) ones. In populations with little or no bird predation located in the middle of the species range the different phenotypes may be selectively equivalent, or nearly so, in which case changes in frequency may be partly due to genetic drift. Indeed, before detailed study of these species in the middle of the twentieth century, all of the variation in phenotype frequency was thought to be due to drift.

From the perspective of evolutionary divergence and speciation, the polymorphisms of shell colour and banding are not the most interesting kinds of variation in these snails. So, let's now ignore them and concentrate on characters that have contributed to the divergence of *C. nemoralis* and *C. hortensis* – in other words characters that help to distinguish them. We will look at three of these characters, which are interestingly different from each other.

First, individuals belonging to *C. nemoralis* almost always have a darkly pigmented lip – the outwardly deflected flange of the shell that forms when the individual reaches adulthood – whereas individuals belonging to *C. hortensis* almost always have a whitish lip. This character appears to be almost completely heritable, with a simple genetic basis. The selective importance of lip colour is unknown. From an ecological perspective, *C. hortensis* tends to be found in cooler and damper habitats than *C. nemoralis*, and its geographic range extends further north (into the southern part of Iceland, where there is no *C. nemoralis*). However, whether a white lip or some character that is correlated with it is selectively advantageous in cooler and/or damper places is an entirely open question.

Second, the shells (and bodies) of *C. nemoralis* adults tend to be larger than those of *C. hortensis*. This time we are dealing with a partially heritable character within each species – both genetic and environmental effects are at work. The heritability of variation in shell size in snails has been estimated to be about 70%. The genetic contribution to shell-size variation is likely to be a large number of QTLs, each with relatively minor effects, as in other cases where the range of phenotypes takes a normal distribution (and in this case just one such distribution, since all individuals are hermaphrodites). The difference between the species is in *average* adult size; the distributions overlap to a considerable extent. Although natural selection often acts on body size, we do not yet have any evidence of it doing so in this case – it is not clear why larger size should have been selectively favoured in *C. nemoralis* or smaller in *C. hortensis*.

Third, the reproductive darts (or 'love-darts') of the two species have different structures. These darts – slender calcareous arrows up to about a centimetre in length in *Cepaea* – are a feature of mating in many species of snails and slugs. Each of the courting individuals fires a dart at its partner at close range during early stages of mating – the dart is fired from an eversible dart-sac. There are different hypotheses about the selective significance of this bizarre system, one of which is that chemical agents carried by the dart enhance the survival of the firer's sperm. Anyhow, in the genus *Cepaea*, *C. nemoralis* has a straight dart with four simple fins, while *C. hortensis* has a slightly curved dart with four bifurcating fins. An extensive study of a large mixed-species colony showed a very low rate of occurrence (less than 1 in 1000) of darts with intermediate structures, suggesting that there are still occasional interspecific matings – unsurprising for a recently separated species pair. The structure of the dart is unlikely to be selectively neutral, given its close association with reproduction; nevertheless the selective significance of the difference between the two species is unknown.

All three of these characters – lip colour, shell size, and dart structure – are developmental in nature. Two of them are specifically adult characters, which develop roughly coincident with sexual maturity: the lip and the dart are absent in juveniles. The third – shell size – is a character that gets finalized at adulthood (about three years of age in *Cepaea*) but has been developing since an early stage. None of the three is associated with any obvious form of developmental constraint or bias. Perhaps, then, this is a case where, in the divergence of the species, we see only the accumulated effects of natural selection and genetic drift.

But perhaps not. This is a good point at which to return to the controversial 1979 paper by Gould and Lewontin in which they described organisms as being 'constrained by phyletic heritage, developmental pathways and general architecture'. Although we can't see any equivalent here of the strange 'odd-only' pattern of centipede segments, there is no doubt that any daughter species that has as its starting point a land-snail ancestor cannot depart too much from the form of that ancestor. For example, a *Cepaea* shell would be much more cryptic in a typical natural habitat with army-style camouflage rather than regular stripes; thus the reason it does not have such camouflage is lack of the necessary variation. At a broader scale, a snail cannot evolve jointed legs or an endoskeleton. This is so obvious that we tend to take it for granted; but maybe we should not. It is another way of saying that too radical a change all at once is either developmentally impossible or is likely to lead to an unfit outcome. Interestingly, then, acknowledging this very general kind of bias is another way of acknowledging the improbability of macromutational change – except for some very specific sorts. One such sort, which we'll look at in the next chapter, is the switch from dextrality to sinistrality (or vice versa) that has happened many times in gastropod evolution – though not in *Cepaea*, where dextrality prevails.

Natural Selection and Developmental Bias

We now turn to one of the best-known examples of what appears to be a developmental constraint or bias – the case of mammalian cervical (neck) vertebrae. These are distinguished from the more posterior thoracic vertebrae by not having attached ribs. The first two cervical vertebrae (C1, C2) are often referred to as the atlas and axis; the others are typically just referred to by their location (C3 onwards). The number of named species of extant mammal is more than 5000 – one 2018 estimate put it at almost 6500, though this is perhaps on the high side. Of these thousands of species, all except a dozen or so have seven cervical vertebrae. The exceptions are as follows: (a) manatees (sea-cows): six cervical vertebrae; (b) two-fingered sloths: five or six cervical vertebrae; (c) three-fingered sloths: eight or nine cervical vertebrae. (Note that all sloths have three toes on their hindlegs, so the commonly used terms 'two-toed sloth' and 'three-toed sloth' are better avoided.)

The first question to ask at this stage is: are these three groups all closely related? The answer is 'no'. Even the two groups of sloths are in different families; and the manatees are in a different order and even superorder from the sloths. Thus each of the three appears to be an independent departure from the 'rule of seven'.

Note that members of the sole group with more than seven neck vertebrae – the three-fingered sloths – do not have particularly long necks. The longest-necked mammal, the giraffe, has the normal complement of seven. In contrast, long-necked birds, such as swans, geese, and flamingos, typically have at least double that number – most species in these groups have between 15 and 25 neck vertebrae. So it looks like there is something different between the developmental systems of birds and mammals that results in the number of neck vertebrae being highly constrained in one group but not in the other. But what is that 'something', and, related to that, what is the nature of the constraint?

This is where we need to ask whether *developmental constraint* and *selective constraint* are different. Since I haven't yet spent much time on the latter term, we need to define it, and we should distinguish these two kinds of constraint – assuming that they really are different. The usual way to distinguish them is to say that in a case of developmental constraint the relevant variants – those needed to evolve in a particular direction – cannot be produced by the developmental system, whereas in a case of selective constraint they can be produced but they are all of lowered fitness. Either of these situations prevents evolution in the direction concerned, but the lack of evolvability occurs for different reasons.

One situation that is clearly selective rather than developmental constraint is when there is stabilizing selection on a continuously variable character such as adult body size. Here, selection favours phenotypes close to the mean; phenotypes further from the mean in both directions are produced by the developmental system, but they do not lead to an evolutionary trend towards larger or smaller body size because they are selected against. In contrast, the reason why there are many animals with legs but none with wheels is due, as noted by Stephen Jay Gould in his 1983 book *Hen's Teeth and Horse's Toes*, to architectural/developmental/functional constraint rather than selective constraint on available variation. Human factories can produce wheels; animal developmental systems probably can't, and even if they could it would be hard to arrange blood and nerve supplies to the wheels. If wheels could be produced and maintained, they might well be selectively favoured.

At this point, the distinction between developmental and selective constraint seems very clear – but is this clarity real, or is it an artefact produced by the contrasting of two rather extreme examples? Let's try to find out, by considering other examples that are somewhere in between the mundane and outlandish ones mentioned in the preceding paragraph.

Why are there no six-legged vertebrates? Is it because the vertebrate developmental system cannot make them, or because they can be made but are unfit? Perhaps vertebrate embryos with six limb buds have been produced from time to time but have always been unfit to the point that they did not survive until birth. Initially, it would seem that if this is the case, then the constraint is selective, whereas if not, and such six-bud variants are never produced, then it is developmental. However, now suppose that variations in early embryonic stages immediately prior to limb-bud initiation include altered patterns of gene expression that would lead to the formation of six limb buds, but that these variants are lethal before the limb buds actually form. This now looks like developmental rather than selective constraint, because the developmental system cannot make six limb buds. But this distinction makes no sense. Can we really say that embryonic lethality (or more generally reduced fitness) before 'stage X' is developmental constraint, while after that stage it is selective constraint?

The problem here is that being unable to make something and being unable to keep it alive are not as different as they initially seem. If animals were made instantaneously, the difference would be real enough. But they're not. Because of development, there is a continuum from one to the other. It is probably the case that no animal developmental trajectories have ever even started off in the direction of producing wheels that rotate on axles – notwithstanding the existence of about 2000 species belonging to the phylum Rotifera ('wheel-bearers', but their circular head structures are not wheels). Thus it seems that wheels cannot be made. Animals that start off making the wrong number of legs are known – e.g. the mutant antennapedia fruit-flies that we saw earlier have eight legs rather than six. These are normally seen as laboratory-produced 'monsters'. They are very unfit and are rarely if ever seen in the wild. Perhaps the lack of eight-legged flies in the wild is due to selective constraint, whereas the lack of six-legged vertebrates is due to developmental constraint?

In my view, the best way to look at this conundrum is as follows. If in a particular group – say at phylum or class level – evolution never, or only very rarely, goes in certain directions, implying a lack of evolvability, this is caused by constraint (*sensu lato*). The constraint might be clearly selective, clearly developmental, or 'in between'. Instances of 'in between' occur where some structure can be made up to a point, in development, but not past that point. In such instances, whether we see developmental or selective constraint depends on which phase of development we examine.

Selective constraint in an early embryo is not like stabilizing selection on adult form. The cause of the selection is unlikely to be differences in susceptibility to some external threat, but rather differences in the integration of the embryo, both in terms of its function at a particular stage and in terms of its ability to develop into later stages. The phrase 'internal selection' is sometimes used to describe selection happening in an embryo for reasons of internal integration, though it is apt to be misleading (selection somehow going on within an individual) and is in my view best avoided.

Having sorted out a few general points, we now return to the question of what causes the lack of evolvability in the number of mammalian cervical vertebrae. There are both developmental and selective aspects of this question. At the developmental level, one hypothesis as to how variant numbers of neck vertebrae arose – for example in two sloth lineages – is that the boundaries of Hox gene expression along the anteroposterior axis became shifted. In particular, Christine Böhmer and her colleagues have suggested, in a paper published in 2018, that the expression patterns of one or two genes of the

Hox-C complex have been shifted anteriorly in the two-fingered sloths and posteriorly in their three-fingered counterparts. This hypothesis needs to be tested by appropriate studies carried out on sloths, but of course such studies are difficult for both practical and ethical reasons – sloths are certainly not 'model animals'.

At the population level, Frietson Galis and her colleagues have championed the hypothesis that a form of selective constraint is involved in stopping most mammals from evolving more or fewer neck vertebrae than the usual seven. In particular, they argue, in papers published in 1999 and 2011, that 'pleiotropic constraint' is at work. Pleiotropy is the production of more than one phenotypic effect by a mutation. It seems that mutations producing a variant number of neck vertebrae in humans are associated with various other developmental changes, including an increase in the occurrence of congenital malformations and early-onset cancers. This association between an altered number of neck vertebrae and other developmental alterations is hypothesized to be due to the lack of modularity at the early developmental stage involved. However, if this hypothesis is correct, it is not clear how sloths (and manatees) manage to break out of the usual constraint.

6 The Evolutionary Origins of Themes and Novelties

Developmental Bias versus Macromutation

In the previous chapter we looked at several different kinds of developmental bias. One of our conclusions was that there are both specific biases, such as the numbers of centipede trunk segments and mammalian neck vertebrae, and general biases, such as the tendency for variant developmental trajectories – and in particular viable ones – to be clustered close to the ancestral trajectory. For example, in the case of snails we noted that the forms of developmental repatterning that were generally available for natural selection to act on were slight quantitative modifications of the pattern of development of the snail that was the ancestor of the clade concerned – an example being developmental trajectories leading to differences in adult shell size. Acknowledging this form of bias entails accepting that evolution of body form does not usually take place via radical-effect macromutations. This is interesting because we saw in Chapter 2 that from the late nineteenth century to the mid-twentieth century, prominent biologists who had a specific interest in the evolution of development, such as William Bateson, D'Arcy Thompson, and Richard Goldschmidt, took a macromutational approach.

D'Arcy Thompson's approach was particularly interesting. As we saw earlier, he argued that the forms of species belonging to different genera or families differed in a coherent way that was amenable to analysis via his geometric transformations. This is a pattern that could be produced gradually by a combination of bias and selection. But he thought that for bigger-scale evolutionary divergences sudden changes were necessary, giving rise to what he called 'new types'. Thus a reasonable interpretation of Thompson's overall view of evolution might be: gradual evolutionary origins of lower taxa such as families; macromutational origins of higher taxa such as phyla. I suspect that this view is wrong, though at our current rather rudimentary stage of knowledge about the origins of body plans, it's hard to be sure. An alternative view is that developmental bias is important in all evolutionary change, and that macromutational repatterning of development contributes only very occasionally to the evolution of body form, with the occasions on which it does contribute being determined by the nature of the macromutation, not by the level of taxon, as we'll see shortly.

The essence of a macromutational change is that it is large enough to be clearly distinct from continuous developmental or phenotypic variation caused by changes in quantitative trait loci (QTLs). Thus while the genetic changes that lead to an increase in shell size over time in a lineage of snails are not macromutational, the genetic change that causes one snail to be different from another in shell colour (e.g. yellow vs. pink) *is* macromutational, in a broad sense of the term. This is ironic, because studies of variation in pigmentation in natural populations of snails and butterflies fall under the heading of ecological genetics, which was one of the main strands of neo-Darwinism for nearly three decades – the 1950s into the 1970s. The practitioners of this discipline were notably anti-Goldschmidt and decidedly gradualist in their view of evolution in general. Yet they worked with macromutations. I think the resolution of this apparent paradox is that they afforded what might be called 'honorary micromutation status' to a genetic change that led, for example, to a snail's shell being pink rather than yellow, or banded rather than unbanded.

Although this at first seems arbitrary and inconsistent, there is a rationale underlying it. Mutations affecting the (2D) pigmentation of an animal are different from those affecting its 3D (or 4D) form. A radical change in colour may well have no knock-on effects elsewhere than the affected body part – in this case the shell. In contrast, a radical change in shape is all-pervasive. Accordingly, we tend to think of the evolution of body shape as being very different in this respect from the evolution of pigmentation. There is, however, one sort of change in shell form (and body form) that has happened in a macromutational manner in gastropod evolution – reversed chirality. This change is of particular interest because it has happened numerous times and it has its basis in a mutation of one of the toolkit genes, as we will now see.

Evolution of Reversed Asymmetry

The Mollusca is the second-largest animal phylum after the Arthropoda. With about 100,000 extant species, there are twice as many kinds of molluscs on the planet as there are kinds of vertebrates. And the vast majority of molluscs (about 80,000) are gastropods. Most gastropods have spiral shells. The exceptions are those slugs (both land and sea) which have completely lost their shells ('semi-slug' is sometimes used for forms with shells that are reduced rather than absent), and those snails that have conical shells (e.g. limpets). Spiral shells can be flat (plano-spiral) as in ramshorn snails, or anything from very squat to very elongate (helico-spiral). All spiral shells coil either to the right or to the left as they grow. The vast majority of spiral snail shells are dextral – that is, they coil to the right. But in a significant minority of cases they are sinistral. The switch from one form of left–right asymmetry to another can only occur 'all at once', since no intermediates are possible.

It is of interest to examine how sinistral forms are distributed across gastropod phylogeny. This is difficult because the conventional phylogeny of gastropods was thrown into disarray by the advent of molecular approaches, and the new, more accurate phylogeny that is taking its place has not yet fully stabilized. However, many families have remained intact, even though higher-level groupings have altered. So, we can at least look at families (and below) in terms of the distribution of chirality.

There are approximately 400 families containing extant gastropods. We can ignore those that do not have spiral shells – though as will become clear later chirality involves asymmetry of the body as well as the shell. Of those families that do have spiral shells, most contain only dextrals. However, there are also families with (a) a minority of sinistral forms (e.g. Partulidae), (b) a majority of sinistral forms (e.g. Clausiliidae), and (c) exclusively sinistral forms (e.g. Planorbidae). An additional complication is that there are a few species, dotted about here and there across the gastropod phylogenetic tree, that have both sinistral and dextral forms. I am not referring here to occasional 'wrong-coil' variants, which can be found in many species. Rather, I am referring to the handful of species in which at least some populations are genuinely polymorphic for chirality, with the rarer shell type (whether dextral or sinistral) making up more than about 10%.

Families that consist largely or wholly of sinistral forms do not make up a tightly knit clade within the gastropods. For example, the two families of this kind mentioned above (Clausiliidae and Planorbidae) belong in different superfamilies and different orders – a conclusion that is unlikely to be altered by further studies on their molecular phylogeny. The existence of wholly sinistral families shows that some instances of chirality reversal in gastropod evolution have led to a large clade of daughter species in which the chirality has then remained unchanged. In other cases, sinistral species appear as rarities in otherwise dextral families; and *Radix peregra* (family Lymnaeidae) provides an example of a variable species with sinistrals as the rarer of the two forms, with only a small number of populations being polymorphic, most 100% dextral.

The various situations regarding chirality that we find in various extant species may represent different points in an evolutionary time series such as the following one:

1. species entirely dextral;
2. species becomes variable (via mutation) with some sinistrals in one population;
3. that population evolves into a daughter species that is entirely sinistral (via selection and reproductive isolation);
4. that sinistral species gives rise to a clade of sinistral daughter species (via speciation unrelated to chirality);
5. a subsequent (much later) speciation event leads to re-reversal and neo-dextrality.

Although the overall pattern of reversals of chirality in gastropod evolution is not yet clear, what is clear is that there have been many such reversals – at least a few tens of them, and perhaps more than 100. Let's now follow up this story about shells with the connecting stories about adult bodies, developmental trajectories, genes involved, and selective scenarios.

If an adult land-snail with a dextral shell emerges sufficiently from its shell, it is possible to see both the respiratory and genital pores (there is just one of each), and both are on the right-hand side of the body. In a sinistral snail, they are both on the left-hand side. These are signs that the snail's body has inverted left–right asymmetry as well as its shell. And naturally this reversed symmetry has its origins in early development. Adult sinistrals develop from juvenile sinistrals. Although it might initially seem likely that, going further back in development, for example to the gastrula, there would be no difference between sinistrals and dextrals, the difference can in fact be traced back to the earliest cleavage divisions of the zygote. The difference between a prospective dextral and a prospective sinistral snail can be seen as early as the eight-cell stage.

The left–right axis is one of the three primary body axes in any bilaterian animal (the others being anteroposterior and dorsoventral). Toolkit genes and cell–cell signalling pathways are responsible for patterning all these axes. With regard to left–right patterning, an important aspect of this, both in molluscs and in vertebrates, is the Nodal signalling pathway. This involves the products of several genes, including *nodal* itself, which encodes a secreted signalling protein belonging to the TGF-beta family (TGF stands for transforming growth factor: 'transforming' refers to cancerous transformation of a cell; a cancerous growth can be seen as development gone wrong). In 2009 it was shown by the biologists Cristina Grande and Nipam Patel that expression patterns of both *nodal* and some genes that are downstream of it in the Nodal signalling pathway, notably *PitX2*, are involved in the reversal of chirality. Interestingly, the inheritance of chirality shows a maternal-effect pattern – that is, the genotype of the mother determines the phenotype of the offspring. This fits with the general pattern that many very early developmental processes are affected by maternal-effect genes in many different kinds of animals. Perhaps the best-known example of this is the role played by the maternally acting *bicoid* gene in the establishment of the anteroposterior axis in *Drosophila*.

So, we have a reasonable understanding of what happens in terms of genetic, developmental, and anatomical changes when a chirality reversal occurs – whether in a speciation event or simply in the transition from parent to offspring in the case of a polymorphic species. However, we do not have as good an understanding of the kind of natural selection that is involved. In fact, it has been demonstrated that, at least in some gastropod species, reversed chirality causes problems for the snails concerned in terms of mating with non-reversed snails belonging to the same species. This represents a form of negative selection acting against snails with reversed coil. When selection on a new variant is negative rather than positive, the expected result is that the gene concerned is removed from the population. So we have a paradox.

The most important work on this topic was carried out on *Partula suturalis* – a species of land-snail that was endemic to the small Pacific island of Moorea, but regrettably became extinct in the wild about 1990 (due to the import and release of a predatory snail from Central America in an abortive attempt at a biological control programme aimed at another alien snail). Before its extinction, *P. suturalis* was composed of dextral populations and sinistral populations, with narrow polymorphic zones between them. The dextral populations of this species corresponded geographically to areas of overlap with populations of sinistral species of *Partula*, as shown by Australian biologist Michael Johnson in 1982. It's known that dextral and sinistral snails have difficulty mating with each other. There is thus a possibility that selection in those areas of overlap favoured dextral individuals of *P. suturalis* in terms of their not interbreeding with the sinistral individuals of the wrong species and thus potentially producing no offspring.

There is, however, a fairly obvious problem with this hypothesis. If a single new phenotypically dextral *P. suturalis* appears in one of these areas, not only will it be at a disadvantage in terms of mating with a member of the wrong species (arguably a good thing), but it will also be at a disadvantage in terms of mating with other members of its own species (definitely a bad thing). The solution to this problem may lie in the maternal effect of the gene concerned. Imagine the situation in which a new mutation (initially invisible phenotypically) produces a mother whose novel genotype codes for dextrality but whose own shell remains sinistral, the same as her parents and grandparents. Such a mother will be able to give rise to a cluster of progeny, some or all of which are dextral. These would be able to mate with each other, though if there is just one such cluster there might be a problem with inbreeding. This is a very different situation from an 'ordinary' (i.e. non-maternal-effect) gene, where a single mutation will only give a single mutant individual. (There is a complication in relation to whether the mutation is genetically dominant or recessive, but that can be ignored here.)

The evolution of reversed asymmetry in gastropods is a fascinating evolutionary story. But what, if anything, does it tell us about the evolution of development more generally? Specifically, is it a model of evolution by macromutation that is the tip of a mostly unseen iceberg? The answer to this latter question is probably 'no'. Rather, I think we should give the same 'honorary micromutation status' to a mutation that reverses the chirality of a snail, just as the ecological geneticists did for a mutation that changes the colour of a shell, for example from yellow to pink. This is because although the mutation leading to reversed chirality is not in the same category as the mutations at QTLs that collectively cause changes in body size, and its effects are very clear and all-pervasive, they are of a rather particular *type*.

When considering all the possible developmental effects of mutations, there are three ways in which we can classify them: by *magnitude* of effect, by *timing* of effect, and by *type* of effect. The 'micromutations' and 'macromutations' that we have been discussing above are categories (with fuzzy boundaries) within the first of these. But using timing and type as well as, or instead of, magnitude of effect is a better way forward. In terms of timing, the evolution of reversed chirality shows that differences between sister-species, and indeed between intraspecific variants, can penetrate back right to the very start of development. This is probably unusual, and therein lies a statistical story that we will return to later.

In terms of *type* of effect on development, a chirality reversal is a very specific thing. We do not yet have a comprehensive catalogue of types of mutational change of development, though we do have various categories. Some changes of a generally 'macro' nature are very different from others – for example, chirality reversal versus homeotic transformation. In his 1986 book *The Blind Watchmaker*, Richard Dawkins distinguished between what we might think of as macromutations *sensu stricto* and *sensu lato*. He calls the former 'true' macromutations. Using an aviation analogy, he exemplifies these by a whirlwind in a scrapyard creating a Boeing 747. In the broader category of macromutations, he also includes the transition between the original Douglas DC8 airliner and its later 'stretched' version with a longer fuselage. Although Dawkins might be described as an arch-gradualist, it is interesting that even he is prepared to acknowledge an evolutionary role for macromutations *sensu lato*.

Finally, we should ask the question: can the reversal of chirality from dextral to sinistral (or vice versa) be reasonably considered to be the origin of an evolutionary novelty? My view is that it cannot. Dextrality and sinistrality can certainly be considered as developmental and morphological 'themes', but the switch from one of these themes to the other is hardly the same as creating, say, a novel structure. Our attention is about to shift to this even more fascinating subject.

The Nature of Novelty

In his 1932 classic *The Causes of Evolution*, the British population geneticist J. B. S. Haldane makes a distinction between routine evolutionary change and the origins of novelties. He puts it as follows:

> The usual course of evolution appears to have been a modification in the relative sizes and shapes of various structures, with very little real novelty.

He goes on to say that novelties *occasionally* arise, and he gives the origins of avian and insect wings as examples.

The origin of evolutionary novelties has become a key focus of evo-devo. So we need to ask how novelties might be defined and classified, and we should examine some relevant case studies. A good launching pad for this endeavour is the 2014 book *Homology, Genes, and Evolutionary Innovation*, by the Yale-based Austrian biologist Günter Wagner, who recognizes two categories of evolutionary novelty, as follows:

Type I novelties – These are novelties that Wagner considers to represent the origin of 'a novel character identity'. Another way of putting this is that the new structure in a descendant species does not have an obvious homologue in the ancestor (or in extant cousins). Wagner gives the insect wing as a paradigmatic example. Another example is the turtle shell.

Type II novelties – In this case, there is an obvious homologue in the ancestor, but it has become radically changed in the descendant. Wagner describes type II novelties as involving 'the evolution of a novel variational modality'. He gives the vertebrate fin-to-limb transition as a paradigmatic example. Other examples include bird and bat wings (modification of ancestral forelegs) and centipede venom claws (also modification of ancestral front legs, but this time one pair out of many as opposed to one pair out of two).

The concept of a variational modality is an interesting one; it can be related to function in the following way. Pectoral and pelvic fins vary enormously across the 30,000 or so species of extant ray-finned fish in size, shape, and position. However, they are generally used for swimming. In some cases they have also acquired other functions – notably for gliding through the air in the case of the flying fishes, with their enlarged, wing-like pectoral fins, and for 'skipping' on quasi-terrestrial surfaces in the case of mudskippers. But they have not lost their main function of propulsion through water. In contrast, when a lineage of lobe-finned fish invaded the land and gave rise to the tetrapod clade, the variational modality was altered from variation among swimming 'designs' to variation among walking 'designs', with some of the latter undergoing another shift in variational modality later in evolutionary time towards propulsion in the air (birds, bats) or back to propulsion in the water (penguins, whales). Wagner comments that shifts such as these between variational modalities are rare, compared to more routine evolutionary changes, and that this implies a degree of 'difficulty' in making them.

There is an interesting connection between this concept of variational modalities and the idea that evolution is often facilitated by organisms having more than one copy of something – whether a gene or a pair of limbs. One copy can more readily evolve a new function if there is at least one other copy that can retain the existing function. With regard to genes, this idea was championed in 1970 by the Japanese geneticist Susumu Ohno in his book *Evolution by Gene Duplication*. With regard to structures, there are many examples, including the various cases of leg 'specialization' and the evolution of vertebrate jaws from gill arches. The bird body plan only works because there was a pair of legs left to walk on land when the other pair became modified to enable flight. However, this is not always the case: none of the three main groups of aquatic mammals has retained rear walking legs, and of course the tetrapods did not retain rear fins. So, as well as there being examples of duplication and divergence, or more generally replication and diversification, there are also cases of evolutionary trends where all replicates shift in parallel.

We will now look at one example each of type I and type II novelties: the shell of turtles and the venom claws of centipedes. As ever, consideration of particular case studies is helpful in understanding the limits of our ability to generalize about evo-devo phenomena. We will end up seeing that both examples have elements of Wagner's type I and type II features. This does not mean that we should abandon the use of these two types of novelty – I think they remain useful – but it does urge appropriate caution in employing them.

Type I Novelty: the Turtle's Shell

There has long been agreement that turtles (*sensu lato*, including the terrestrial tortoises) make up a monophyletic group – i.e. a clade. The clade consists of about 300 species, and is usually given the taxonomic rank of an order, namely Testudines. However, where this clade falls in relation to the rest of the reptiles has been much debated. Early phylogenetic trees of the tetrapods based on morphological characters often put the turtles as a basally branching group, one that split off from other reptiles before various groups of those (e.g. lizards, crocodiles) began to diverge from each other. However, several molecular phylogenetic studies have now been carried out and a consensus is emerging that turtles are not basally branching after all. One specific hypothesis for the phylogenetic placement of turtles is that they are the sister-group to the archosaurs – the group that includes crocodiles, dinosaurs, and birds (Figure 6.1).

The turtle shell is made up of a dorsal part, often dome-shaped, called the carapace, and a ventral part, generally much flatter, called the plastron. These two parts are mostly fused together at the sides. The only openings are the holes where the head and limbs can be extended and retracted. At first sight, the turtle's shell has no homologue in any other group of tetrapods. This would seem to make it a classic type I novelty. However, discoveries of various fossil forms over the last couple of decades have shown that the situation is more complex than was previously thought.

Fossil turtles of the genus *Odontochelys* from the mid-Triassic period, about 220 million years ago (mya), seem to have had a plastron but not a carapace. And an earlier fossil called *Pappochelys*, from the early Triassic, about 240 mya, had a series of dermal 'belly bones' called gastralia. These may have collectively been a forerunner of the plastron, or at least of part of it. If this is the case, then Wagner's criterion of a type I novelty not having a homologue in an ancestor may not apply to the plastron. However, the carapace might yet turn out not to have a homologous structure in ancestors. The carapace evolved quite soon after the plastron in geological terms – the first completely shelled turtle, *Proganochelys*, has been dated to about 210 mya.

So, it looks like the turtle's shell evolved in stages, over millions of years. The ecological context for those stages isn't clear because we don't know for sure whether the earliest turtles (*sensu lato*) were marine, freshwater, coastal, or

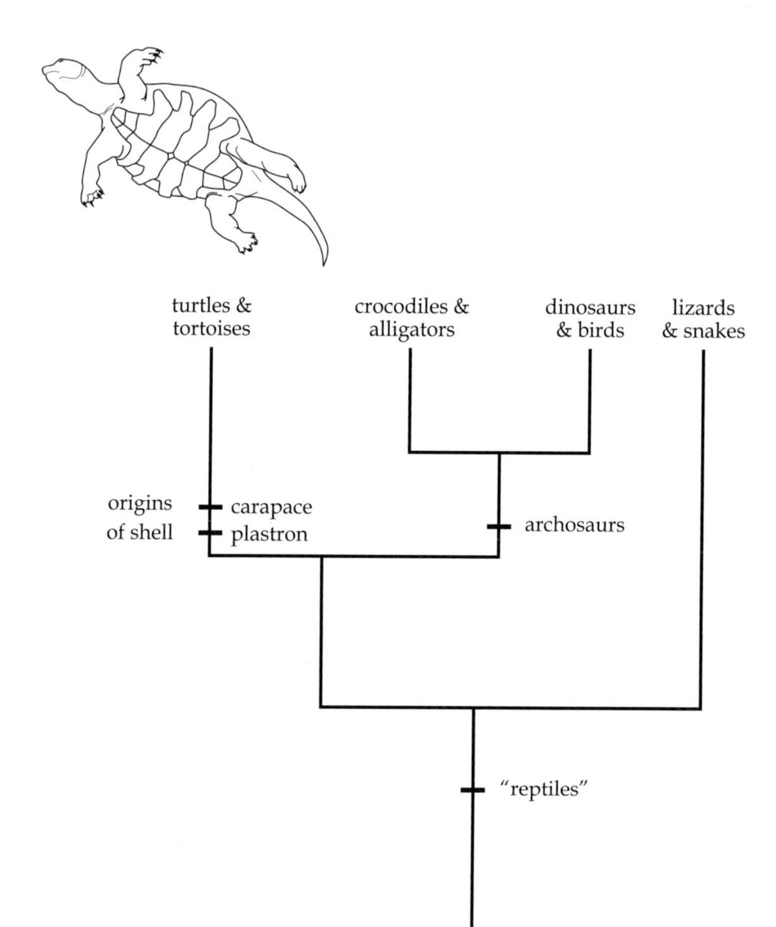

Figure 6.1 One hypothesis of the evolutionary relationship between the turtles and other groups of 'reptiles' ('Reptilia' is no longer regarded as a valid formal group name, because it is not a clade). The shell arose in the turtle stem lineage, as shown, in stages rather than 'all at once'. Fossil evidence (see text) suggests that the ventral part of the shell (the plastron) evolved before the dorsal part (the carapace).

terrestrial. There is even a theory that some of them had a burrowing habit, and thus were at least partly subterranean. But whatever the ecological context, and whatever the exact selective reasons, it is clear that the evolutionary trajectory from unshelled to shelled tetrapod involved radical changes in body form, not the simple addition of a shell to an otherwise unchanged architecture.

These radical changes include the following. First, in development, the ribs do not grow significantly both outward (laterally) and downward (ventrally) as they do in most tetrapods. Rather, they grow mostly outward. So they do not meet at the breastbone and enclose the trunk in a rib-cage as they do in most reptiles and in other tetrapods – birds and mammals. There is no breastbone in a turtle. The protective job of the ventral ribs and breastbone is done instead by the plastron. Second, the ribs end up being fused into the carapace. Third, because of this, the shoulder blade (scapula) cannot be outside the ribs as it is in nearly all extant tetrapods – if it was, it would have to be outside the shell too, which would clearly be a non-functional arrangement.

The turtle's internal scapula, as contrasted with an external one like our own (which we can easily feel the position of) initially seems like a binary switch – akin perhaps to that between dextral and sinistral gastropods. If this were true, then involvement of a large-effect mutation in a developmental gene might be suspected. But it's not true. A gastropod shell is a single integral hard part – a tube that is coiled on itself. In contrast, the carapace, plastron, spine, and shoulder girdle are all made up of multiple bony units. Also, as well as potentially being inside or outside the ribs, the shoulder girdle can be anterior to them. This opens up various evolutionary possibilities, including the girdle going gradually from external to internal via anterior, or indeed vice versa.

To make some headway in understanding the evolutionary novelty of the turtle's shell, it helps to focus not on the shoulder girdle but on the ribs, and in particular on how these are 'persuaded' to grow outward but not (much) downward in the embryo. It turns out that the key embryonic structures in this respect are the carapacial ridges. There is one of these running along each side of the turtle embryo. These ridges secrete the protein products of key toolkit genes that attract the growing embryonic rib-precursors in a lateral direction. Grace Loredo and colleagues showed in 2001 that one of these is a protein called FGF-10 (FGF stands for fibroblast growth factor; a fibroblast is a type of cell). To connect this with our gene classification of Chapter 4, look at Figure 4.2 and find Hedgehog. The FGF proteins are like the Hedgehog protein in their role – they are mobile ligands involved in cell–cell signalling, so they fit into the category 'other-6'.

Although FGF-10 is involved in developmental processes in other vertebrates, there is no obvious equivalent of the carapacial ridge in any other vertebrate groups. So the origin of the turtle shell as a structure that characterizes adults (and indeed the majority of the life cycle) involved the origin of this smaller-scale embryonic structure. That constitutes a partial explanation – but only that. Where did the carapacial ridge come from, and what were the factors driving its evolutionary appearance? At this level, our explanation still falls short.

Type II Novelty: the Centipede's Venom Claws

Like turtles, centipedes form a well-defined clade, one which is usually given the taxonomic rank of a class, namely Chilopoda. In this case, the clade consists of a few thousand extant species (about 4000 are known at present) rather than a few hundred. The phylogenetic relationships among the constituent orders of the class are well established; and the placement of centipedes against the broader backcloth of myriapod and arthropod phylogeny is now reasonably certain – see Figure 6.2. The venom claws – also called forcipules – are unique to centipedes, and must have originated as a novelty in the centipede stem lineage. There are no known descendant species that have lost them. The claws are a robust pair of appendages on the first post-cephalic segment, in other words the segment that is behind the head but in front of the first leg-bearing segment. They are used to inject venom into prey, which is a wide range of invertebrates and, in the case of the large tropical species, small vertebrates too.

It seems likely that the pair of venom claws represents an evolutionary modification of what was the first pair of walking legs in an ancient ancestor. Exactly how ancient is hard to say, because the centipede fossil record is rather poor. However, it is clear that some, and perhaps all, of the centipede orders had come into existence by the late Carboniferous, before 300 mya. Fossil forms belonging to two of the four main orders are found in the Mazon Creek fossil beds, which are located in Grundy County, Illinois, a short

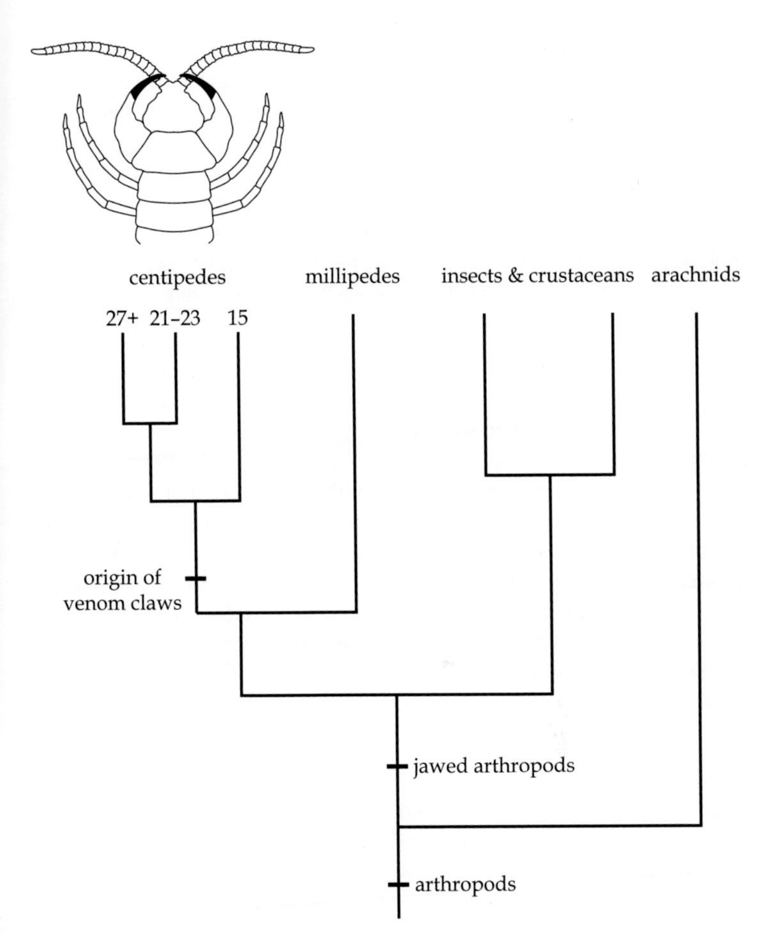

Figure 6.2 The evolutionary position of centipedes in relation to other myriapods (e.g. millipedes), and other arthropods more generally. The venom claws evolved in the centipede stem lineage. They are possessed by all known centipedes and by no other arthropods. Although some other arthropod groups use venom, their delivery systems for it are independent evolutionary inventions, not homologues of centipede forcipules. The classification of centipedes is slightly simplified to show just three groups, distinguished on the basis of their numbers of leg-bearing segments.

distance to the southwest of Chicago. And the centipede stem lineage must have existed long before that, because there are Devonian fossils (*c.* 380 mya) belonging to an extinct order of centipedes – appropriately called the Devonobiomorpha. There are fossil legs that have been interpreted as having belonged to centipedes from around 420 mya (the boundary between the Silurian and Devonian periods). And there are fossil millipedes from the Silurian. So it seems likely that the split between centipedes and millipedes took place sometime in the early Silurian, perhaps about 440 mya. In general, myriapods are thought to have been among the first land animals.

The hypothesis that the venom claws are evolutionarily modified legs is supported by the finding that the segment on which they are borne (the forcipular segment) is characterized by expression of the Hox gene *antennapedia*, which is normally considered to be a trunk gene rather than a head gene. This finding was reported by American biologists Cynthia Hughes and Thomas Kaufman in 2002. In the modification of a pair of legs into a pair of venom claws, several things happened. They can broadly be grouped into changes in external form and changes in internal anatomy. Externally, venom claws are typically thicker than legs, more heavily armoured, and tapering to a stout tip. Internally, in addition to changes in the musculature, repatterning of development has produced a venom sac – usually at the base of the claw – and a venom duct leading from this to a position near the tip of the claw, where there is an opening through which the venom leaves the centipede and enters the body of its prey.

The way in which a venom claw develops in a present-day centipede suggests a hypothesis as to how claws evolved from legs. Work published in 2012 by the French biologist Michel Dugon and his colleagues has shown, in a species of the genus *Scolopendra*, that the venom duct arises from an invagination of the cuticle, running along the length of the claw from near its base to near its tip. This invagination eventually becomes a tube that detaches itself from the cuticle and becomes an internal structure, with the venom gland at its basal end. It is possible that the first stage in the evolution of venom claws was secretion of venom from glands opening onto the surface of the first pair of legs – these perhaps having arisen by ectopic expression of genes involved in glandular systems elsewhere in the body. The next stage may have been a channel along the outside of the incipient claw, allowing more targeted flow of venom. The final stage would have been the internalization of the channel to form a duct.

The external changes in shape that led from a leg to a venom claw are easy to envisage as having taken place gradually over a long period of evolutionary time. Indeed, the most basally branching centipedes – the long-legged scutigeromorphs – exhibit a slender type of venom claw that can be thought of as being morphologically intermediate between the original first leg and the chunkier claws of the more derived orders. So, from the perspective of external shape, the venom claw represents a classic type II novelty. However, from the perspective of the internal anatomy, and in particular the venom sac and duct, this novelty is harder to classify. There are no obvious homologues of these structures in other groups of myriapods, or indeed in arthropods more generally. If such homologues are genuinely lacking, then from an internal perspective the venom claw should be described as a type I rather than a type II novelty.

From Case Studies to Generalizations

In this chapter, we have looked at three examples of evolutionary origins: sinistral snails (multiple clades of various sizes, up to the family level); turtles, with their unique shells (one clade, about 300 species); and centipedes, with their venom claws (one clade, about 4000 species). All these new layouts of part or all of the body of the animals concerned can be described as new evolutionary themes. And except for the sinistral snails they can reasonably be described as evolutionary novelties. In each case, we examined the phylogenetic placement of the forms concerned against the backcloth of the broader group inside which the clades examined fell. We considered the likely time of origin of these clades, in terms of geological periods. We also looked at the nature of the developmental repatterning that took place in each, and some of the developmental genes involved. Having done that, it is time to stand well back from our three trees and try to glimpse an outline of the whole wood. At the current limited stage in our knowledge of the origins of new themes/novelties, this is necessarily a somewhat subjective endeavour. However, it is well worth undertaking, and as an aid to this exercise I would recommend Lewis Held's 2014 book *How the Snake Lost its Legs*, which deals with many additional case studies of the origins of themes and novelties.

Here, then, are a few subjective views on possible generalizations about these origins.

First, novelties and themes can originate at any point in evolutionary history. There was not a novelty-generating period followed by a period of merely quantitative modification of earlier-established themes. (Whether this is also true in relation to body plans we will consider in the next chapter.) Centipede venom claws originated in the Palaeozoic era, the turtle shell in the Mesozoic. The direction of coiling of gastropod shells has altered many times during the evolution of this class of molluscs, probably from shortly after their origin in the early Palaeozoic (or possibly even earlier) right up to the late Cenozoic. The origins of new themes, including those that can be described as type I and type II novelties, occur only rarely compared with the ubiquitous routine evolution of characters such as body size, but their occurrence is scattered in geological time, not clustered into particular eras or periods.

Second, although it is often hard to separate the origin of type I novelties from their type II counterparts, the latter type of novelty is clearly the commoner of the two. In contrast, type I novelties are comparatively rare. When Wagner distinguished between types I and II, he used the insect wing as a paradigmatic example of type I, as we noted earlier; this is appropriate, because this type of wing did not evolve from a leg, as those of birds did. He also gave a second paradigmatic example – the vertebrate head. Again, this is appropriate, because the closest vertebrate relatives are headless. Lancelets and sea-squirts have no head, so our own head seems to have been added after the divergence of these close relatives.

Third, novelties are often but not always associated with invasion of a new type of habitat – especially in relation to the three major habitat types of land, water, and air. Examples include all types of wings in the case of groups becoming aerial (birds, bats, insects), walking legs in groups invading the land from the sea (tetrapods, various arthropod groups), and fins or flippers in the case of groups going back to the water (penguins, whales, seals, sea-cows). In other cases, novelties are associated with a particular mode of life rather than a particular habitat. The venom claws of centipedes are associated with a predatory lifestyle, and have been retained in all descendants of the centipede stem lineage regardless of whether they are soil-burrowers, surface-dwellers, tree-climbers, or coastal species. The turtle shell is also lifestyle-related, but this time in connection with defence rather than attack. Again, the shell has persisted in all descendant forms, regardless of whether they are marine, freshwater, or terrestrial. In a few groups the shell has become less protective – for example in the leatherback, where instead of the carapace being bony it is made of hardened skin. Perhaps the sheer size of this species renders protection less crucial to it than to others.

Since we will shortly be moving on to consider body plans, the relationship between novelties and body plans deserves brief consideration here. The turtle's shell is a novelty that arose in a particular lineage of vertebrates and represents a notable modification of the vertebrate body plan. Likewise, the insect wing is a novelty that arose in a particular lineage of arthropods and represents a notable modification of their body plan. The tentacles of cephalopods – which have not been mentioned up to now – represent a notable modification of the molluscan body plan. And so on. Thus in a sense novelties are nested within body plans. In a similar way, lesser themes, such as sinistrality in gastropods, are nested under novelties. And routine evolutionary changes are nested under themes. In each case, a modification of the lower level of change leaves the higher one in place. Perhaps this hierarchical picture of evolution as a whole is telling us something important.

7 The Evolutionary Origins of Body Plans

The Pattern of Animal Relatedness

It is constructive to approach this issue from a historical perspective. Some aspects of animal relatedness have been known for a long time – centuries – while some have only been established in the last few decades. And others remain to be worked out or confirmed. A useful starting point for this historical approach is the 1817 four-volume work *Le Règne Animal* (The Animal Kingdom) by the French comparative anatomist Georges Cuvier, who divided the kingdom into four *embranchements* (branches): vertebrates, molluscs, articulates (outwardly segmented animals), and radiates (radially symmetrical animals). We should note here that Cuvier was an anti-evolutionist; he was opposed to the evolutionary theories of his fellow Frenchmen Jean-Baptiste Lamarck and Étienne Geoffroy Saint-Hilaire, and he did not live to see the publication of Darwin's *Origin of Species*. However, many non-evolutionists prior to Darwin (from Aristotle onwards) made good attempts at the classification of animals, even though the fruits of their labours would not be given an evolutionary interpretation until later. Here, I will discuss Cuvier's suggested groups as being evolutionary ones, even though that is not how he saw them.

The relatedness of all vertebrates that Cuvier considered clear in 1817 is equally clear now, more than two centuries later. The Vertebrata is a well-established clade, usually thought of as a sub-phylum of the Chordata. The apparent relatedness of all articulates – the arthropods and annelids (segmented worms) – seemed clear enough to most biologists well into the twentieth century, even though it turned out to be wrong,. In 1932, J. B. S. Haldane wrote 'there can be little doubt that arthropods are descended from annelid worms'; and the British zoologist Robert Clark, in his 1964 book *Dynamics in Metazoan Evolution*, refers to 'Annelids and their undisputed relatives, the Arthropoda'. It was only in 1997, as we noted in Chapter 3, that molecular phylogenetic studies showed that annelids and arthropods are not sister-groups, but in fact rather distant evolutionary cousins. Cuvier's 'molluscs' grouping has remained largely intact, though two groups that he included there have had to be removed – barnacles are arthropods, and brachiopods (alias lamp shells) constitute a phylum in their own right. However, while Cuvier's grouping together of the various classes of molluscs – notably gastropods, bivalves, and cephalopods – has stood the test of time, the subgroups that he recognized within the Gastropoda have not survived later analyses.

It is probably true to say that the grouping of animals into phyla (and sub-phyla, where these apply) has been relatively stable for the last half-century. In many cases, the division of phyla into classes has also been stable, and in some cases the division of classes into orders has been stable too. What has been – and in some cases still is – more changeable is the way in which the groupings at any one of these levels are related to each other. For example, the perceived pattern of relatedness among the mammalian orders has altered in recent years, as has the perceived pattern of relatedness among arthropod sub-phyla, with insects now being seen to be more closely related to crustaceans than they are to myriapods (centipedes and millipedes). And the relatedness of phyla in terms of super-phyletic groupings has perhaps changed the most. Hemichordates, despite their name, are more closely related to echinoderms than to chordates; and the relationships among the basal animal phyla (Placozoa, Porifera, Cnidaria, and Ctenophora) continue to be uncertain.

Given this state of affairs, we might reasonably ask: is it possible to give a picture of high-level animal relationships, in particular relationships among phyla and sub-phyla (since these are the groupings to which 'body plans' most readily apply), that is likely to withstand the test of time? I think that it's finally possible to give a broad outline of super-phyletic relationships that is not likely to change – or is only likely to change minimally – as future research into these issues continues. This outline is shown in Figure 7.1. As can be seen, it gives only seven phyla (out of approximately 35), but the affiliations of many others are given in the legend. Three main features of this branching

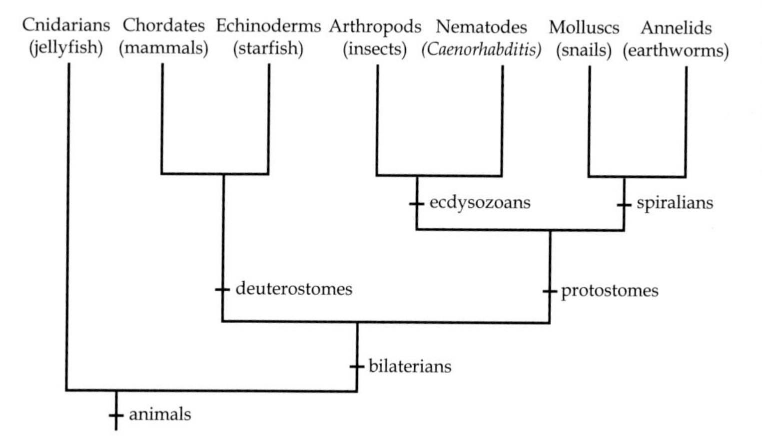

Figure 7.1 High-level relationships within the animal kingdom, shown by focusing on seven of the best-known phyla, each of whose names is followed by an example of an especially well-known subgroup. Note the following key super-groups, which are described further in the text: Bilateria, Protostomia, Deuterostomia, Ecdysozoa, and Spiralia. Here are where some of the less well-known phyla fall, with a bias towards those that receive at least a brief mention elsewhere in this book. Deuterostomia: includes hemichordates. Ecdysozoa: includes the 'worm' phyla of priapulids and nematomorphs, also the 'pan-arthropod' phyla of tardigrades and onychophorans. Spiralia: includes the brachiopods (lamp shells), rotifers (wheel animals), and bryozoans (moss animals). Basal animals (branching off early): include Porifera (sponges), Ctenophora (comb jellies), and Placozoa (flat animals).

diagram, and the distribution of species across its various branches, are as follows:

1. The super-phyletic grouping of Bilateria – animals that are bilaterally symmetrical and/or had bilaterally symmetrical ancestors – comprises the vast majority of the animal kingdom (31 out of 35 phyla; more than 98% of extant animal species).
2. Bilaterian phyla can be placed into one or other of the super-groups protostomes (mouth-first in embryogenesis) and deuterostomes (mouth-second).
3. The protostome super-group has two major subdivisions – the Ecdysozoa (named after growth by moulting) and the Spiralia (named after a spiral

pattern of embryonic cleavage), within one or other of which all protostome phyla can be placed.

The body plan of any phylum or sub-phylum can be described by a small set of key features that, together, distinguish it from any other body plan. For example, if an animal is externally segmented and has an exoskeleton it is an arthropod. If, on the other hand, it is internally segmented and has an endoskeleton it is a vertebrate. If it has a mantle and a shell (external, internal, or vestigial) it is a mollusc. And so on. We can often describe each body plan more fully by adding descriptors that apply at a higher level – for example the arthropod and vertebrate body plans can be described as also being bilaterally symmetrical. However, this doesn't always work. For example, echinoderms are part of the Bilateria, but the body plan of present-day adult echinoderms is not bilaterally symmetrical, due to important developmental repatterning events that occurred early in the history of the phylum.

So, body plans characterize higher taxa, notably phyla or sub-phyla. Any animal can be placed in a particular high-level taxon by examination of its body plan. Each body plan is characterized by a set of features; and each body plan is distinct from each other one. These facts naturally lead to questions about when and how body plans originated in evolution. We'll deal with the former question first as it is the easier of the two – but still by no means easy.

The Timescale of Body-Plan Origins

The seven phyla used in Figure 7.1 to show the broad outline of the structure of the animal kingdom were chosen on three bases. First, they are reasonably well known (in contrast, for example, to the phyla Kinorhyncha and Phoronida – google them if you wish!). Second, and related to that, they contain large numbers of species (thousands). Third, all or some of their members have structures that fossilize easily – vertebrate bones for example (Chordata), or the hard parts of corals (Cnidaria, though note that other cnidarians, such as jellyfish, have no hard parts). But there is one exception to this last criterion – nematodes. These are soft-bodied worms that rarely fossilize.

It is almost certain that all the phyla shown in Figure 7.1 were in existence by the later stages of the Cambrian period (see geological timescale of Figure 7.2), in other words by about 500 million years ago (mya). How

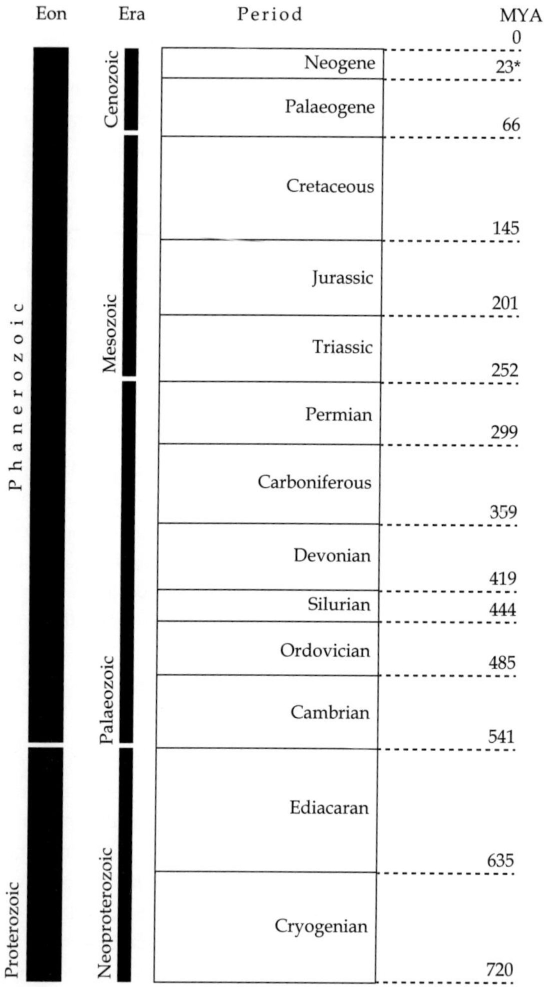

Eon	Era	Period	MYA
			0
Phanerozoic	Cenozoic	Neogene	23*
		Palaeogene	66
	Mesozoic	Cretaceous	145
		Jurassic	201
		Triassic	252
	Palaeozoic	Permian	299
		Carboniferous	359
		Devonian	419
		Silurian	444
		Ordovician	485
		Cambrian	541
Proterozoic	Neoproterozoic	Ediacaran	635
		Cryogenian	720

Figure 7.2 Geological timescale, extending back to the period before the first animals are thought to have evolved. The broadest division is into eons, which are divided into eras, and these in turn are divided into periods. Periods can be further subdivided into epochs, but here I do not go into that level of detail. * Last 2.5 million years = Quaternary.

much earlier than that they arose is a moot point, which we'll get to shortly. The important thing, for now, is that since basal (non-bilaterian) animals and all super-groups of bilaterians were in existence in the Cambrian, it seems likely that most or all animal phyla were in existence then. Thus the origin of animal body plans is an issue that concerns the Cambrian period – or before. This makes the origins of body plans very different from the origins of novelties, which, as we saw in the previous chapter, can originate at any time in animal evolution (e.g. centipede venom claws in the post-Cambrian part of the Palaeozoic, turtle shells much later, in the Mesozoic). But as ever we should be wary of sharp distinctions. For example, the origin of vertebrate jaws, which we looked at in Chapter 3 as an example of 'general adaptation', was a major modification of the vertebrate body plan, and probably happened in stages during the Ordovician and Silurian periods.

It is often said that the majority of animal body plans arose in what is called the Cambrian explosion. However, there is much uncertainty about the nature of this 'explosion' of animal forms, and whether it did indeed take place in the Cambrian (541 to 485 mya), or earlier – perhaps in the Ediacaran (635 to 541 mya).

Many fossils from the Cambrian can be clearly assigned to phyla that are used to group present-day animals. For example, trilobites are clearly arthropods. Several less-well-known fossils (such as *Pikaia* from the Burgess Shale in British Columbia) are clearly chordates. There was a Cambrian group of shelled animals called helcionellids, which were almost certainly molluscs. Other Cambrian fossils are more enigmatic, but nevertheless seem to fit into one of the recognized animal super-groups – for example a scaly creature about five millimetres long called *Wiwaxia* is certainly an animal, probably a spiralian, and possibly a mollusc.

The situation changes completely when we go back to the Ediacaran period. There are no obvious animals in this period, even the later parts of it as we near the base of the Cambrian (say about 550 mya). A debate has been ongoing ever since the discovery of the Ediacaran biota about its phylogenetic status. Some palaeontologists think that many of the Ediacaran organisms are animals and belong in certain animal phyla – including Cnidaria, Mollusca, and Annelida. Others think that they belonged to a separate kingdom of multicellular organisms – one that is now extinct – and that they were not animals at all; this idea was championed by the German palaeontologist Dolf Seilacher. And of course a third possibility is that some of the Ediacarans were one of these and some were the other. The continuing uncertainty is nicely illustrated by the Ediacaran fossil called *Kimberella*. Some palaeontologists see it as a mollusc, others as a bilaterian animal that cannot be assigned to any particular phylum, and others as a multicellular organism of uncertain kingdom.

The continuing uncertainty over the nature of the Ediacaran organisms means that we do not know whether the 'Cambrian explosion' of animal form really happened early in the Cambrian or at some stage in the Ediacaran. But what about the question of whether it might have been earlier still? Before the Ediacaran was the Cryogenian period (720 to 635 mya), whose name speaks for itself. This is the period that some scientists call 'snowball Earth'. There were major worldwide glaciations through most of the Cryogenian – the Sturtian glaciation followed by the Marinoan. It seems a very unlikely period for the animal kingdom to have got started. There are no certain animal fossils from this period, and indeed very few fossils that anyone could reasonably suggest even *might* have been animals. Some early 'molecular clock' studies of the animal kingdom suggested that animals originated in the Cryogenian or even earlier, but it now seems likely that these studies were subject to various flaws. More recent molecular estimates of the time of origin of the animal kingdom are more in line with estimates based on fossils.

So, it seems that animal body plans originated sometime between about 630 and 530 mya. They may have evolved gradually across that vast 100-million-year timespan, or much more rapidly, say from 540 to 530 mya – a mere 10-million-year period. The American palaeontologist James Valentine favoured the latter sort of timespan. Others who favour a lengthier period use phrases such as 'slow-burn' (as opposed to explosion) and 'ghost lineages'. The idea of the latter phrase is that a lineage of animals may exist over a long period of time – many millions of years – without leaving any fossils at all. This might occur because of (a) poor conditions for fossilization during the period concerned, or (b) a lack of readily fossilizable features and/or small size of the animals concerned, or (c) subduction of the relevant sedimentary rocks by subsequent plate movements.

A specific question that helps to illustrate this general issue is: why are there no trilobite fossils from before about 520–530 mya? Perhaps trilobites were too small and/or insufficiently armoured to leave fossils, or perhaps they simply did not exist. While this question may seem rather vague and speculative, it relates to a testable hypothesis. If a single Ediacaran fossil trilobite is ever discovered, the hypothesis that there were no pre-Cambrian members of this group will be conclusively disproved.

Mechanisms of Body-Plan Origins

In the present section we start by focusing on population-level mechanisms and end up thinking in terms of developmental stages. In the next section, we start with developmental stages and end up with genes, looked at from a general perspective. In the following chapter, we look at the roles of some specific genes (or gene groups) in producing certain body-plan features.

Starting with populations, then, we need to think in terms of developmental variation being present in an Ediacaran or early Cambrian population of a species of animal that will end up being ancestral to a large group of descendants. How might variant developmental trajectories spread in such a species, both within and between populations? The first thing we can say is that variants having an effect on the overall body layout are unlikely to be selectively neutral. We should perhaps recall (from Chapter 1) Brian Goodwin's view that this is a possibility, and that in such a situation the direction of evolution might be determined entirely by developmental bias; but we should also remind ourselves that this seems very unlikely. Assuming, then, that selection *was* involved, it probably operated in much the same way as it does in natural populations today. This statement links to the 'uniformitarian' world-view, namely the idea that, in general, natural processes operating in the past worked in much the same way as they do today, a world-view that was adopted by the Scottish geologists James Hutton and Charles Lyell and named by the English polymath William Whewell (who also coined the term scientist). Uniformitarianism is a good starting point for thinking about the distant past – in a sense it keeps us 'grounded'. However, we need to keep in mind that it may turn out to be wrong.

What *form* of selection was involved in the spread of variants contributing ultimately to the establishment of a new body plan? Well, directional rather than stabilizing, to be sure. And probably of the sort that produces general rather than special adaptation, in the terms we discussed in Chapter 3. But beyond those rather general claims it is hard to go. To go further we would need to know the particular starting point and endpoint. For example, the original bilaterian animal (the 'urbilaterian', as it is often called) was probably produced by modification of a radially symmetrical animal, something broadly resembling a cnidarian (e.g. jellyfish) or a ctenophore (comb-jelly). A change in symmetry such as this might have involved a certain kind of macromutation (*sensu lato*), just as evolutionary reversals in the left–right asymmetry of a gastropod do. However, going from radial to bilateral is a much bigger step than going from sinistral to dextral, and thus carries with it a greater concern about possible fitness decrease if it happened all at once. So perhaps the origin of bilateral symmetry happened in stages – though what they might have been remains a mystery.

The stages in the evolution of some other body-plan features are clearer. For example, the evolution from an appendage-free vermiform plan to a form with articulated appendages might have proceeded via rudimentary un-jointed appendages, and may have involved a gradual lengthening of them. Indeed, the lobopod design, as seen in onychophorans (velvet worms), with its stumpy lobe-like legs that lack joints, is often thought of as having been a transitional stage in the evolution of the arthropod body plan, with its more sophisticated jointed legs and other appendages.

We don't know much about the genes that might transform a lobopod leg into a jointed one, but we do know something about the genes that were involved in a later stage in the elaboration of the arthropod body plan, namely the reduction of the number of pairs of legs from 'many' to three in the origin of the insects. In 1978, the American geneticist Ed Lewis published a paper that, as we saw in Chapter 1, was very influential in the early days of evo-devo. In it, he notes that insects came from many-legged ancestors – though he took on board the then-prevalent idea that those ancestors were millipedes, whereas we now know that they were crustaceans. He states that the origin of insects must have involved the evolution of 'leg-suppressing' genes, such as some of the Hox genes. Wisely, however, Lewis did not spell out a particular mechanism for this to happen, which is what had led to the rejection of Richard Goldschmidt's proposal of evolution of the arthropod body plan and its major variants via 'hopeful monsters'. In contrast, Lewis left open the possibility of a succession of individually small-effect mutations in what are usually thought of as large-effect genes.

In between the origin of a bilaterian body plan and the origin of the more specific arthropod body plan, one of the important evolutionary events that took place was the split between protostomes and deuterostomes. As noted in Chapter 2, a deuterostome such as a vertebrate can be pictured as an upside-down version of a protostome such as an arthropod; the first person who drew attention to this was Étienne Geoffroy Saint-Hilaire in 1822. The main nerve cord is ventral (and often paired) in arthropods, dorsal in vertebrates. Relative to the nerve cord, the digestive tract is ventral in vertebrates, dorsal in arthropods. Such a situation could have been produced by an evolutionary axis reversal in the stem lineage of one or other of these two super-groups, in which case we are back to changes in body symmetry, and hence possibly back to macromutation (*sensu lato*). However, another possibility is that both body layouts arose from a last common ancestor that had only a diffuse nerve network. In this case, both the protostome ventral nerve cord and the deuterostome dorsal one (lost in echinoderms) might have evolved gradually from the pre-existing diffuse network. To know which of these alternatives is correct, we would need to know the body plan of the urbilaterian animal, which may never be possible. However, this has not stopped people from suggesting possible forms for 'Urby' (as it is often nicknamed) – we will consider these in the next chapter.

The above considerations suggest that we should not rule out a role for macromutations (in the broad sense) in the origins of animal body plans. However, we should not assume their importance either. Indeed, now is the time to move towards a more sophisticated view than micromutation versus macromutation. In doing this, we need to (a) acknowledge that 'magnitude of effect' of a mutation is a continuum rather than a dichotomy and (b) bring developmental time into the picture. The former is easier than the latter, so we'll start there.

The terms micromutation and macromutation simply refer to opposite ends of a continuum of the magnitude of phenotypic effect that a mutation can have. Analysis of the effects of QTLs on continuously variable traits such as body size and shape show that some 'micromutations' have bigger effects than others, so there is no clearly defined entity called a micromutation. Equally, as we saw in Chapter 6, Richard Dawkins had to resort to using 'true macro-mutations' to distinguish some big-effect mutations from others – the creation of a Boeing 747 from scrap metal being a 'true' one but the stretching of a

Douglas DC8 into a longer version of the same airliner being implicitly an 'untrue' one. In discussing macromutations in the previous and present chapters I have resorted to a broadly similar tactic, with large-effect mutations on shell pigmentation and chirality being included in macromutation *sensu lato*, but not in the narrower version (*sensu stricto*) which only includes mutations like antennapedia.

When the British geneticist Ronald Fisher articulated the case against involvement of macromutations in evolution, he used an abstract continuum of 'magnitude of change' (Figure 7.3). He employed a simple logical argument to show that as the magnitude of a mutation's phenotypic change increased, its probability of being selectively advantageous decreased – ultimately to zero, though interestingly Fisher qualifies this by saying 'or at least negligible'. The difference between these two – zero and negligible – is huge when we are dealing with timespans of many millions of years. While this is an important criticism of Fisher's approach, there is another even more important one: Fisher portrays 'magnitude of change' as a *single abstract dimension*. In a real organism that is built by real developmental processes, there are many types of change – i.e. many types of developmental repatterning. Type and magnitude are inextricably linked. Considered from a developmental perspective, 'magnitude' is a complex multidimensional concept, not a simple unidimensional one.

Evo-devo has not yet reached a point where the intertwining of magnitude and type of change in development can be dealt with in a comprehensive way. However, we can at least inject one very real dimension into the idea of generalized magnitude of effect – the point in developmental time at which ontogenetic trajectories begin to diverge as a result of a mutation in a developmental gene. Such a mutation can cause the new trajectory to diverge from the original one at any point from zygote onwards, depending on when the gene is first expressed. If we compare two direct-developing species within a genus, it is usually – but not always – the case that their developmental trajectories do not begin to diverge until quite late in embryogenesis. However, if we compare two species from different phyla, the trajectories either diverge right from the start or from shortly after that.

Contrasting the typical developmental divergence of animals representing closely related species with animals representing the different body plans that characterize phyla and sub-phyla, as in the previous paragraph, is a good

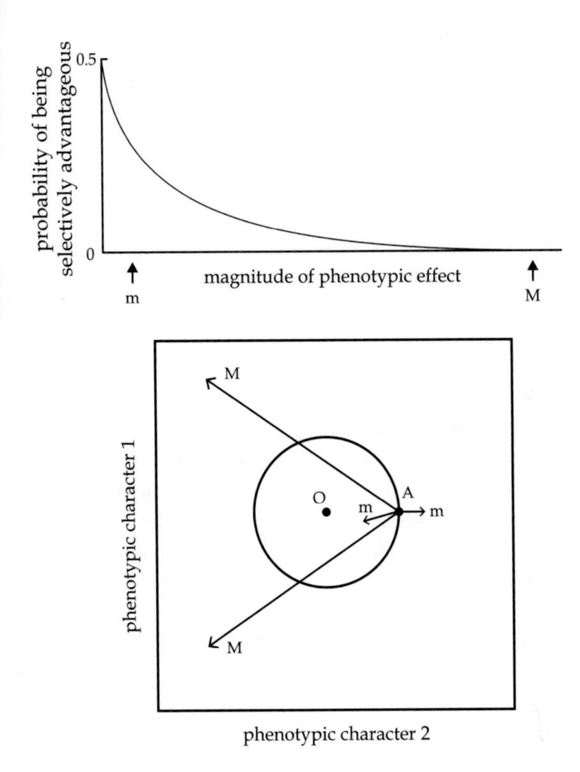

Figure 7.3 Ronald Fisher's argument for a negative relationship between the magnitude of a mutation's phenotypic effect and its probability of being selectively advantageous. Top: the proposed relationship, in which the probability of being advantageous declines from about 0.5 for very small-effect mutations to about zero for very large-effect ones. Bottom: the logical basis for the argument, in which the axes are the values of two (unspecified) phenotypic characters. O represents the optimum joint phenotype and A the actual one at a given moment in time. Starting from A, a small-effect mutation m has a reasonable chance of being advantageous. In contrast, a large-effect mutation M has a zero chance if it shifts the joint character value by an amount that is greater than the diameter of the circle centred on O. For criticism of this argument, see text.

starting point for trying to understand whether there is something special about how body plans arise. But there are many levels in the taxonomic hierarchy between species and phyla, and we need to take these into account. There is nothing magic about the phylum level of organization. It does not constitute some mystical realm in which processes fundamentally different to those of 'routine' evolution take place. Rather, it represents one end of what is probably best described as a continuum. But a continuum of what? Well, both a continuum of 'body-plan-ness' and a continuum of earliness of developmental divergence.

As we saw in Chapter 3, the overall network of molecular and cellular interactions that constitutes the development of a particular type of animal includes many different kinds of pattern, one of which – an important one – is a hierarchy. In fact, development as a whole can be seen as a hierarchical pattern of the splitting of cell lineages, which is in some cases known in considerable detail – for example in the roundworm model system *Caenorhabditis* that we examined in that chapter. The pattern of cell lineage splitting can be thought of as the outcome of a network of causal links which itself also has a hierarchical component (as well as other components – combinatorial ones, feedback ones, and so on). So, as well as thinking of the living world as a taxonomic or evolutionary hierarchy, we can think of the ontogeny of a single animal as a developmental or morphogenetic hierarchy. This naturally raises the question of what is the relationship between the two.

A good starting point for addressing this question can be found in *Morphogenesis and Evolution*, a book published in 1988 by the British biologist Keith Thomson. In Chapter 7, Thomson makes the following claim:

> In principle we should be able to reconstruct for any species or any higher group a sequence of levels of morphological characteristics that define all the higher groups to which the taxon belongs, and to match these up with particular points in the hierarchy of morphogenesis.

This statement can be taken as being representative of a general approach to the relationship between evolution and development that can be dubbed the 'hierarchy correspondence' approach, which several other authors have taken, including the Austrian biologist Rupert Riedl in his 1978 book *Order in Living Organisms*, and myself in 1984, in *Mechanisms of Morphological Evolution*. This is not to say that our views were the same – they weren't – but they have much in common.

Thomson's statement is very abstract. It often helps if we think in terms of a concrete example. Let's use the mouse model system to exemplify what

Thomson means. The species *Mus musculus* belongs to the order Rodentia, which in turn belongs to the class Mammalia. Mammals belong to the vertebrate sub-phylum of the phylum Chordata. That's the taxonomic hierarchy side of things. Now we turn to the 'sequence of morphological characters' that Thomson refers to. From the top down, in terms of the taxonomic hierarchy, characters that put mice in a particular taxon at each level are as follows:

1. The (embryonic) notochord shows that the mouse is a chordate.
2. Its endoskeleton shows that it's a vertebrate.
3. Its fur and its nipples show that it's a mammal.
4. Its incisor teeth show that it's a rodent.

The morphological characters that distinguish a mouse from other rodents, especially closely related ones such as other mice and rats, are subtler, and include many quantitative characters such as body size and shape.

Note that Thomson's 'sequence of morphological characters' was just the first step from the taxonomic towards the developmental hierarchy. Now we need to make the second step – from morphological characters to morphogenesis. Here, developmental time becomes an explicit variable, and we can examine the stage of development at which each character originates, as follows:

1. The notochord that identifies the mouse as a chordate is formed early during embryogenesis.
2. The skeleton that identifies it as a vertebrate begins to form later – but still quite early.
3. Hair follicles, indicating mammalian identity, begin to develop at a stage of embryogenesis that is later again.
4. The characteristic rodent incisor teeth develop later still.

Many of the quantitative characters that distinguish mice from rats do not get finalized until late post-embryonic development, in some cases approaching adulthood.

So, there is good evidence for a relationship between the taxonomic and developmental hierarchies in this example. And it is easy to find other examples in which this relationship also exists. However, in all cases it is a statistical relationship rather than an absolute one. For example, between the class Mammalia and the sub-phylum Vertebrata, mice belong to the clade Amniota, which includes all mammals, birds, and reptiles, but excludes amphibians and fish. The defining characteristic of this clade is, of course, the amniotic egg, which makes its appearance at the start of the life cycle (or at the end, taking a different perspective). It does not fall in between the origin of the skeleton and that of hair follicles. For those who like nice neat correspondences, this is a pity. However, for those who acknowledge that evolution is a messy process, a statistical rather than an absolute correspondence between the taxonomic and developmental hierarchies is the best that could be hoped for, and is very interesting in its own right.

So far, we have considered the relationship between the two hierarchies from the perspective of a single species; now we need to take a comparative approach. Initially, this takes us back to von Baer and his pattern of embryonic divergence. For example, mouse and dog developmental trajectories start rather similar but begin to diverge visibly at a certain stage, somewhat later than the pharyngula, and after this they become progressively more different from each other. This divergent pattern is captured in von Baer's laws, which we looked at in Chapter 2. So far so good; we acknowledge that often there is an 'hourglass complication', but otherwise the essentially divergent nature of the developmental trajectories is clear. However, we must now extend the picture into another dimension that was not explicitly covered by von Baer's laws: taxonomic breadth of comparison.

Consider a comparison involving a mouse and a snake. This is a comparison between classes of vertebrates rather than between families of mammals and hence is taxonomically broader: mouse and snake embryos diverge earlier than mouse and dog embryos. Part of this difference is due to hourglass effects – snakes have eggs that are very different from those of mammals. However, if we restrict our attention to the pharyngula and subsequent stages, the pattern is still evident: mice and snakes diverge before mice and dogs. The conclusion to this line of reasoning is that the 'depth' to which evolutionary changes in development penetrate back towards the start of the developmental process increases with breadth of taxonomic comparison.

How does such a pattern involving developmental time and taxonomic breadth come into existence via accumulated evolutionary change over many millions of years, in multiple lineages that stemmed from an ancient common ancestor and diverged from that ancestor (and each other) to varying degrees? The most likely scenario is that lineage divergences that lead to new higher taxa are based not on typical speciation events (such as those that lead to the splitting of a species of finch) but on atypical ones – those few that involve changes in earlier developmental stages than usual.

Genes and Body-Plan Origins

If the evolutionary divergence of body plans differs statistically from the divergence of closely related species in terms of the stages of development that are typically affected, is it true that these two types of evolutionary divergence also differ in the developmental genes that are typically involved? The answer to this question would be straightforward if there was a clear correspondence between genes and developmental stages, but there is not. We know this for at least two reasons: first, because some minor-effect mutations of toolkit genes are contributors to routine microevolutionary changes that involve late-stage developmental repatterning; and second, because toolkit genes may be expressed for long or multiple periods during development.

Let's look at two papers that help to expand the latter point. The title of a 1996 paper by the American biologists S. J. Salser and Cynthia Kenyon expresses it beautifully: 'A *C. elegans* Hox gene switches on, off, on and off again to regulate proliferation, differentiation and morphogenesis.' So does the title of a 1998 paper by the British biologist Michael Akam: 'Hox genes: from master genes to micromanagers.' Akam notes that the *Drosophila* Hox gene *Ultrabithorax* (*Ubx*) 'micromanages segment development by manipulating a large number of different targets at many developmental stages'. He goes on to say that the genes targeted by *Ubx* are not all just next-level-down genes ('sub-masters'), but rather that they are varied types of gene and are controlled by *Ubx* 'throughout development'.

So, evolutionary changes affecting early and late development need not involve different genes; they may instead involve different mutations of the same genes. For example, mutations in certain Hox genes can affect the early-stage development of limbs, while different mutations of those same Hox genes can affect the identities of the digits that form at later stages of limb development. However, we should now switch back from specific examples involving Hox genes to the general picture. Let's approach it in the following way. Suppose we look at all the genes that are expressed (switched on) at two stages of development, one early and one late, in a particular animal – whether a worm, a fly, a mouse, or a human. To what extent do the sets of all expressed genes at the two stages chosen overlap with each other? To give a specific answer to this question – for example 42% – we would have to specify the animal and the exact two stages of development. But that's not necessary if we just want a general answer, and here it is: the overlap is partial, it is rarely or never either total or zero. It logically follows from this general answer that evolutionary changes involving 'stage a' in development may involve the same or different genes as those involving 'stage b'.

Here is a recap of the main messages of the last two sections. The population-level processes that are involved in the origins of body plans may not be qualitatively different from those that lead to typical speciation events; certainly, both involve natural selection. And the genes involved in the two sorts of divergence may also not be different. But two other things *are* different. First, the developmental stages are different, though, as we have seen, in a statistical rather than absolute way. Second, while the genes may not be different, the mutations of them that characterize the two kinds of divergence are different (again statistically rather than in absolute terms); for example, the two groups of mutations may affect regulatory regions of the gene that function at different developmental stages. Finally, it is better not to think just about the two extreme kinds of divergence – those that lead to the divergence of similar daughter species and those that lead ultimately to body-plan divergence – but rather to think of them in a broader context – the way in which the evolutionary and developmental hierarchies are related overall.

8 Body-Plan Features and Toolkit Genes

Homologous versus Convergent Features

Body-plan features that have been discussed so far include symmetry, segmentation, skeletons, and limbs. When these are encountered in different phyla, are they homologous or convergent? There are examples of both of these, plus examples where the answer is not yet clear. Bilateral symmetry of the overall body plan seems to have originated just once. So the fact that vertebrates and arthropods are both bilaterally symmetrical is due to their having inherited that body layout from their last common ancestor; in other words, their bilaterality is homologous. However, although vertebrates and arthropods both have skeletons (whereas animals belonging to many other phyla do not) these represent convergent rather than homologous skeletons – this is clear from the fact that one is 'endo', the other 'exo'. Turning to segments and limbs, the fact that both vertebrates and arthropods have these component parts is hard to interpret with certainty one way or the other. The reason for this is our lack of knowledge of that ancient animal that we call the urbilaterian, or 'Urby' for short. Direct evidence of this creature will probably never be forthcoming, since it was almost certainly small and soft-bodied, and has left us with no fossils from which to infer its living form. Instead, we can only make rather indirect inferences based on the point in the animal evolutionary tree at which we think bilaterality arose. However, indirect inference is better than nothing, so here goes.

The urbilaterian was an aquatic animal. We know this because the fauna of the Ediacaran and Cambrian periods was entirely aquatic. The fossil record suggests that the first animals to invade the land – arthropods – did so in the

Silurian period (about 444 to 419 mya – see Figure 7.2), with the first land vertebrates following in the Devonian (419 to 359 mya). Of course, the earliest known fossils of a group of animals only provide a latest possible origin of that group; the actual origin could have been – and almost certainly was – much earlier. This means that the very first animals to invade the land might conceivably have lived in Ordovician times (485 to 444 mya), but the probability of them having existed earlier still is negligible. So the urbilaterian, which lived more than (and probably *much* more than) 540 mya, was aquatic, and it was almost certainly marine. This animal must have arisen as an offshoot from one of the basal non-bilaterian groups, most of which are exclusively marine, with the fourth (Cnidaria) being overwhelmingly marine and with just a few freshwater representatives (e.g. *Hydra*).

A plausible scenario (but only that) for the evolution of the urbilaterian is that it arose as an offshoot from one of the asymmetric or radially symmetrical groups in an Ediacaran ocean. In contrast to the sedentary or swimming lifestyle of those groups (e.g. sedentary sponges or swimming jellyfish), Urby moved slowly across the sea-floor in much the same way as some present-day free-living flatworms do. The evolutionary changes that gave rise to it were probably driven by selection for this mode of existence, though as noted in the previous chapter we do not yet know how gradual this process was, nor via what intermediate stages radial symmetry might have given way to its bilateral counterpart.

Given that early radially symmetrical animals were of relatively simple construction, it seems reasonable to believe that the urbilaterian was simple too. It probably had few cell types and few organs. It was probably unsegmented and without limbs. It might have had light-sensitive pigment spots, but not complex eyes. This view of the nature of Urby is called the 'simple urbilaterian' hypothesis, as promoted in 2008 by Andreas Hejnol and Mark Martindale. There is, however, a contrasting 'complex urbilaterian' hypothesis, suggesting that Urby had most or all of the features that the other hypothesis says it lacks. One version of this – the segmented urbilaterian – was proposed in a 2003 paper by French biologists Guillaume Balavoine and André Adoutte. The idea of a simple rather than complex urbilaterian is the more parsimonious, and accordingly it gets my support. However, not only might this hypothesis be wrong, but in fact the two hypotheses merely represent opposite ends of a continuum of possible levels of complexity that this early animal may have had.

The importance of this issue is that in order to know which features of later major groups of animals (such as arthropods and vertebrates) are homologous and which are convergent, we need to know if they were present, at least in rudimentary form, in their last common ancestor. But wait a moment: the urbilaterian was not that ancestor. This point needs careful thought, and in particular careful phylogenetic thought. 'Urbilaterian' is the name we give to the *first* animal that was bilaterally symmetrical. From the urbilaterian stem arose most of the phyla of the animal kingdom – about 30 (out of 35) of them. But they did not radiate instantaneously. Rather, the process of their radiation took many millions of years – we do not know exactly how many million. The lineage that split into the two main groups of bilaterians – protostomes and deuterostomes – was the last common ancestor (LCA) of the arthropods and the vertebrates. We know that this LCA (let's call it 'Lasty') lived after the urbilaterian, but not by how long. How similar the two animals – Urby and Lasty – were is not known. It is possible that the latter was more complex than the former, but if so, by how much? There are clearly many unanswerable questions here.

Given these uncertainties over the nature of the last common ancestor of the bilaterian animal phyla, it is hard to know if body-plan features that we observe in two or more of these 30 or so phyla are homologous or convergent. What are the main features at issue here? Whether the body plan is segmental or not is a key feature. If a body is built in a segmental manner, this has an all-pervasive effect. Segmentation, once evolved, is difficult to lose. In the three great groups of segmented animals – arthropods, annelid worms, and chordates – there are relatively few cases of loss of adult segmentation (more in the annelids than the other two phyla), and arguably no complete losses of all traces of segmentation throughout the entire life cycle.

Other important features are more easily lost. These include limbs and eyes. In the vertebrates, snakes illustrate loss of the former, blind cave-fish loss of the latter. In the arthropods, parasitic barnacles illustrate the loss of limbs (in adults), soil-dwelling centipedes the loss of eyes. However, it is always hard to be definitive about what body-plan features can or cannot be lost. For all bilaterian phyla, the fundamental bilateral symmetry would seem to be the most deep-seated architectural feature – deeper even than being either segmented or unsegmented. Yet echinoderms have dispensed with it (except in their larval stages), and in doing so have dispensed with brains, among other things. Also, it is always important to remember that our vantage point on evolution is somewhere in the middle of the process – we do not know what evolution will do in the future.

In three of the sections of this chapter we will look at three different body-plan features – specifically segments, limbs, and eyes. For each of these, we will focus on a gene, or a group of genes, that is involved in the development of the feature concerned. In each case we will ask the question: are these homologous genes driving the development of homologous or convergent features, and if the latter, how does this situation arise? But before we do this, we need to delve further into three interrelated things: the toolkit genes themselves (building on what we learned in Chapter 4); the question of how general we should expect their activities to be across the animal kingdom; and the question of how general and specific mechanisms interact to produce biological diversity.

Genes and Generality

Genetics and zoology are very different undertakings. Genetics emphasizes generality, zoology emphasizes diversity. When James Watson and Francis Crick published their famous 1953 paper on the double-helical structure of DNA, they did not mention a single taxonomic group. That's hardly surprising, because it was assumed that the structure of DNA applied across the living world, an assumption that has been largely borne out. In contrast, important discoveries in zoology are usually much more taxon-specific. For example, the discovery of living coelacanths – a type of fish that is closely related to the land vertebrates and that had been thought to be extinct for millions of years – was published by the South African scientist J. L. B. Smith in 1939. His paper was published in the same scientific journal as Watson and Crick's paper – *Nature* – and the gap between them in time was a mere 14 years. Both papers were commendably short, as papers in *Nature* often are. But there the similarity ended. While Smith could claim to be reporting the zoological discovery of the century – it was often reported as that – it was a very taxon-specific one. For students of other animal groups than the vertebrates it had no particular import. It was a discovery of a particular animal form, and forms belong to the realm of diversity. But the genes that generate forms belong to the realm of generality. Somehow, in getting from genes to animal structures via developmental processes, we move from the general to the particular. How this happens, and how the way in which it happens gets modified in evolution, is the stuff of evo-devo, and the toolkit genes are key players.

Let's probe the business of getting from the general to the particular. Not only is the double-helical structure of DNA a generality of the living world, but so too is the genetic code – the way in which genes make proteins. There is a little variation in this, but not much. For example, mitochondrial genomes often use a slightly different code than nuclear genomes. And nuclear genes in a few types of organism – e.g. some yeasts – use a code that is slightly different from the nuclear-gene norm. Given the near-universality of the genetic code, the structure of the coding region of a gene is much the same across the living world in general terms (readable in triplets), though it is interrupted by many non-coding stretches (introns, as we saw in Chapter 1) in some cases and few or none in others. However, although the code is near-universal, the actual DNA sequence of the coding region of one gene is different from that of another. This is where the general begins to give way to the particular. Also, even if two genes encode very similar proteins, they may have different regulatory regions so that the proteins are produced in different parts of the embryo or at different stages of development.

We now move on to the question of the breadth of taxon to which a particular class of gene relates. This is very variable indeed. At one extreme, all organisms have genes for standard metabolic enzymes such as those that conduct the energy-releasing process of glycolysis (breakdown of glucose). At the other, there are some proteins that are only found in particular taxa and are absent from all others. Some groups of fish that live in sub-zero waters in the Arctic and Antarctic have proteins that bind to carbohydrates to form a sort of antifreeze. These proteins – and the genes that encode them – are absent from other fish, and indeed from most other groups of animals. The DNA sequences of these genes in cold-adapted fish suggest that the antifreeze proteins evolved from gut enzymes. This serves to remind us that, just as all animals are related, all genes are related. When we talk about a particular group of genes – such as the Hox genes – we must always remember that the group concerned is part of a hierarchy of relatedness just as a group of animals is. For example, Hox genes are just one subset of homeobox-containing genes, as we saw in Figure 4.2. Equally (but not shown in that figure), starting with Hox genes and going down rather than up in the hierarchy of genes, we find subgroups and sub-subgroups that can be distinguished from each other in terms of degree of sequence similarity.

Before the discovery of homeobox genes in the early 1980s, most biologists suspected that while many of the genes for metabolic enzymes might be similar across the living world, the genes that are responsible for making parts of bodies (then unknown) would be very different from group to group, simply because the structures they make are so different. For example, it was not really expected that the same genes would turn out to be involved in making vertebrate and arthropod legs. Nor was it expected that the genes responsible for making the complex eyes of vertebrates would be the same as those involved in the convergent complex eyes of cephalopods, still less that those same genes would be involved in building the differently structured compound eyes of insects. And yet we now know that these things are true – for example, genes of the Pax6 family regulate the making of eyes throughout the Bilateria.

Another way to put the pre-homeobox expectation is this. In those days, most biologists who thought about this issue imagined that the *same genes* – in the sense of members of the same gene family – were involved in the development of homologous structures (e.g. the legs of a mouse and those of a dog), but *completely different genes* were responsible for the development of convergent structures. We now know that the latter is untrue. However, the situation is complicated by the difficulty often encountered in deciding whether structures are homologous in the first place, as we saw in the previous section. If the last common ancestor of arthropods and vertebrates did not have legs, then the legs of these two groups are convergent. But what if the ancestor had tiny appendages of some sort? Perhaps then arthropod and vertebrate appendages are homologous but their legs are not? This would be no different from saying that bird and bat forelimbs are homologous while their wings are not, which is known to be true. So, in each of the following three sections we will examine the degree to which it is possible to draw conclusions about the homologous or convergent origin of the structures concerned.

This story takes us to the concept of 'deep homology', which has become something of a buzzword in evo-devo. The 'deep' means deep in evolutionary time. There is now known to be deep homology of the toolkit genes. For example, arthropod and vertebrate Hox genes are homologous, despite the very different morphology of the animals concerned. Indeed, all animal Hox genes are homologous. But to what extent does the concept of deep homology extend to the structures of animals? By the end of the next three sections, we should be able to appreciate how much is known about the answer to this fascinating question.

Segments, Notch Signalling, and Hox Genes

Vertebrates, arthropods, and annelid worms all have segmental body plans. We have already seen that, contrary to earlier beliefs, arthropods and annelids are only very distantly related. And both of these groups are even more distantly related to the vertebrates than they are to each other. Close relatives of all three groups are unsegmented. These facts suggest that segmentation arose independently in each of them. And the variation in the mode of development among the three groups, including the different roles of mesoderm and ectoderm, also suggest independent origins. If this 'independence hypothesis' is correct (which goes hand-in-hand with the simple urbilaterian hypothesis), we might not expect similar genes and signalling pathways to be involved in the generation of segments in the three groups. However, this turns out not to be true, as we will shortly see.

Comparing the three big segmented groups is complicated by the fact that the way in which segments are produced developmentally *within* each of them can vary. This is unsurprising, especially in the arthropods, where the number of species involved is more than a million. The other two groups are also quite large, with about 50,000 species of vertebrates and 25,000 of annelids. The main arthropod model system – *Drosophila* – is rather atypical for arthropods in general, because of the near-simultaneous formation of all of its various segments along the anteroposterior axis. This is a feature only of the group of insects described as 'long germ', meaning the whole length of the embryo at the germ-band stage is segmented at about the same time. For short-germ insects and other groups of arthropods, the norm is for segments to be generated in a sequential manner from head to tail. And this is also how segments develop in both vertebrates and annelids.

An aspect of the development of segments that is shared between the three groups is Notch signalling. Recall from Chapters 3 and 4 that there are a few key signalling pathways that are involved in development across the animal kingdom. So far, we have only discussed the Hedgehog pathway, but now it is time to discuss its Notch counterpart. Like Hedgehog, Notch follows the broad pattern shown for a generic signalling pathway in Figure 3.3, but with some exceptions. For example, in Notch signalling, both the ligand and the receptor are membrane-bound. Thus this pathway only works between cells that are juxtaposed – hence Notch signalling is often referred to, for example in Scott Gilbert's 2016 blockbuster textbook *Developmental Biology*, as a juxtacrine system.

Although some other aspects of the segmentation processes in vertebrates, arthropods, and annelids (inasmuch as we can generalize about them) appear to be unique to one or another of these three groups, the discovery that Notch signalling is an important underlying similarity among them takes us back to the question of whether the three segmentation processes are homologous or convergent. There is not yet unanimity on this issue; and indeed it is complicated by the fact that in between the whole animal being segmented or unsegmented are intermediate arrangements such as some organ systems being segmented and others not. However, there seems to be an increasingly persuasive case that genes and signalling pathways that were already involved in the elongation of the anteroposterior axis that characterizes the development of all bilaterian animals – including Notch signalling – were recruited to help organize the segmentation process as it evolved. Such recruitment of genes and their products to a novel (albeit related) function, argued for by Ariel Chipman in 2010, can be referred to as gene co-option. It appears to be a very important evolutionary process. It will turn up again in the other examples of body-plan features that follow; and we will look at it in general terms in the concluding section of this chapter.

So far, the segmentation story as I have told it concerns the making of segments, and one particular end result of this process, namely segment number (regardless of segment type). This number may be fixed or variable within a species. We have already seen a contrast of this kind within the arthropods, with insects usually being fixed, some groups of centipede variable. In vertebrates, frog species have fixed segment numbers while some species of snakes have variable ones. In general, it is probably the case that a small number of segments can be produced repeatably, without any variation, while a large number cannot.

But there is another side to segmentation than controlling the process of producing segmental blocks of tissue and determining how many there are. This takes us to the topic of segment identity, which we met in Chapters 1 and 4. As we noted there, the Hox genes are segment-identity genes. Different *types* of segment in vertebrates, arthropods, and annelids (i.e. segments with different identities) are characterized by the expression of different combinations of Hox genes. The way in which the pattern of Hox gene expression varies going from head to tail is broadly similar in the three groups. Thus, in terms of determining segment identity, the similarity in the developmental genetics of segmentation among the three groups is perhaps even greater than in relation to segment number. The interpretation of this finding is not simple. However, as with genes determining segment number, those determining segment identity may have acquired their roles in segmentation via co-option.

Many aspects of the complex process of segmentation have been glossed over in the above account. The number of segments varies enormously among different members of the three segmented groups of animals. It's typically fewer than 50 (e.g. in humans, insects, and leeches) but it can be more than 100 (e.g. some snakes, millipedes, and polychaete worms). Also, segmentation happens against the background of both direct and indirect developmental systems. All three groups vary in this respect. There are many other complications of segmentation that could be discussed, but I don't think that any of them detract from the general message of the importance of gene co-option in the evolutionary repatterning of development from an unsegmented to a segmented body plan.

Limbs, *distal-less*, and Dlx Genes

A broadly similar evo-devo story to the one about segments can be told about limbs. My starting point for this is a mutant fly. Back in the mid-1970s, when I was a PhD student studying laboratory populations of *Drosophila* fruit-flies, one of the many mutant forms that I came across was distal-less, an almost self-explanatory term: the distal ends of the legs are missing. Like most other large-effect mutations of developmental genes in *Drosophila*, this one is detrimental to the fly – it is better to have complete legs than truncated ones.

So it does not tell us anything about evolution. But it most certainly does tell us something about development, and this developmental message is much clearer now than it was in the 1970s.

Back then, mutations were generally discovered via, and called after, their visible mutant effects. We're already familiar with these in terms of homeotic mutants such as antennapedia, where what is visible is a pair of legs growing out from the head instead of antennae. But how the genes produced these effects was largely unknown in the 1970s. However, with the flowering of both developmental genetics and evo-devo in the 1980s, the missing mechanisms began to be seen. It turns out that in a normal fly the *distal-less* gene is expressed in the distal parts of the legs – the parts that fail to form when the gene is unable to function. This makes perfect sense. Further, it turns out that *distal-less* is a homeobox gene. In other words, it makes a transcription factor that turns downstream genes on. These genes, or a subset of them, are responsible for building the tip of the leg.

Sequencing genes in *Drosophila* allows them to be searched for in other animals; and the *distal-less* gene, or more precisely homologues of it, turns up all over the place. They are found throughout the animal kingdom. In vertebrates, there is a family of distal-less genes called the Dlx family.

Although, like most toolkit genes, *distal-less* has more than one developmental role (we saw an example of this in Chapter 5 with butterflies), a common finding across many animal groups is that *distal-less* is expressed at the tips of developing limbs, just as it is in flies. For example, it is expressed in mouse limb buds, showing a parallel between vertebrates and arthropods. In terms of its expression in other animal phyla, the American biologist Grace Panganiban and her colleagues showed in 1997 that distal-less genes are expressed in developing appendages in annelid worms, velvet worms (onychophorans), sea-squirts, and echinoderms. These last animals are particularly interesting in this respect, because they have such an unusual body plan as non-bilaterally symmetrical bilaterians. Picture a starfish. It has a number of 'arms' (usually but not always five); growing out from under these arms are hundreds of tiny 'tube-feet'. It is at the tips of growing tube-feet that *distal-less* expression is found.

It's clear that echinoderm tube-feet are not homologous to arthropod or vertebrate limbs – and indeed these two types of limbs are probably not homologous to each other. As we have seen, the last common ancestor of arthropods and vertebrates was probably a rather simple marine flatworm lacking features such as limbs or segments. However, that does not mean that it necessarily lacked some simple appendages, perhaps small sensory outgrowths. If it had such outgrowths, then appendages in a very general sense are homologous between arthropods and vertebrates, and perhaps across many bilaterian phyla. If it did not, then it must nevertheless have had a distal-less gene, performing some other function. In the former case, we would not describe the situation as gene co-option, because the gene already had the appendage role in the ancestor, but in the latter case we would describe it as co-option, because the gene has been co-opted for the new role independently in different descendant groups. Either way, there is deep homology of the gene. And in the second scenario there is also (arguably) deep homology of appendages – but not of limbs, as already noted.

In this context, it is worth noting the different usages of these various terms for things that stick out from the main body. Appendage is the broadest term – it can refer to *anything* that sticks out, with the proviso that it's multicellular, including, for example, antennae. Limb is more restricted – it refers to appendages that are used for locomotion, though generally it excludes fins, which is why we sometimes talk about the vertebrate fin-to-limb transition. However, it does not exclude all appendages used for movement in aquatic habitats – for example, trilobites had limbs rather than fins. My *Concise Oxford Dictionary* defines limb as meaning 'arm, leg, or wing'. Hence 'leg' is narrower still – it clearly refers to motion on a solid surface and does not include wings, though again the solid surface may be the base of an aquatic habitat – so a trilobite's limbs can also be described as legs. However, the further we extend our usage of these terms into the realm of *all* the animal phyla, with their varied kinds of appendages, the harder it becomes to be certain about the best usage.

Eyes, *eyeless*, and Pax Genes

The story of eye evolution has much in common with the stories of segments and limbs, as told above, and again it involves gene co-option. Once more, the issue of definition is important: what is an eye? At one level, it is a complex organ capable of producing detailed visual images. At another, it is a device for detecting different levels of light, and might be something as simple as a pigment spot. In between these two extremes there are 'simple eyes' called ocelli, found in a wide range of animal groups – including insects, which have compound eyes as well. The camera-type eyes of vertebrates and cephalopods are famous for their close resemblance, despite their origin by convergent evolution, which is given away by the fact that our eyes have a 'blind spot' while theirs do not. Outside the Bilateria, box jellyfish have remarkably complex eyes despite the fact that they have no brain with which to process the visual information that the eyes gather – an interesting evolutionary puzzle.

As with limbs, the eye story can be started with a mutant fruit-fly: eyeless. This parallels distal-less; but note that there was no 'segment-less' mutant to start the first of the three stories. This fact helps to emphasize the point that segmentation is in some sense a more deeply seated aspect of the architecture of an animal than are limbs or eyes. A segmented animal cannot completely dispense with segments. There are mutant forms of *Drosophila* larvae that lack some of their segments, but none that lack segments altogether. And those that lack blocks of segments – the 'gap mutants' – do not progress very far developmentally; rather, they die at an early stage.

Like Hox genes and *distal-less*, the *eyeless* gene makes a transcription factor. The vertebrate homologue of insect *eyeless* is called *Pax6*. This name contains two pieces of information. First, 'Pax' denotes a conserved sequence in the gene called a 'paired box', which is a bit like the homeobox sequence in that it encodes a DNA-binding domain in the resultant protein. However, compared to the homeodomain's 60 amino acids, the paired domain is more than twice as long, with 128 of them. Second, the '6' indicates that there is a family of Pax genes in vertebrates. (Just to complicate matters, some Pax genes, including *Pax6*, have a homeobox as well as a paired box, but that's another story.)

Homologues of insect *eyeless* and vertebrate *Pax6* genes have been discovered to be involved in the development of eyes almost wherever they are found in the Bilateria, as noted in a 2005 review by Zbynek Kozmik. In basally branching animal groups, ancestral genes of the Pax family are found, but not *Pax6*. The eyes of box jellies show expression of a gene called *PaxB*. This gene is also found in sponges, which lack eyes entirely. So the family of genes is older than what we might call the 'family of eyes'. Pax genes have probably never been lost in animal evolution – partly due to the usual multi-functionality of toolkit genes, which means that genes tend to be conserved even if one of their functions is lost. The eye function most certainly is lost in various lineages, since eyes themselves – like limbs – are often lost. So far, we have noted the loss of eyes in some cave-dwelling fish and soil-dwelling centipedes. Eyes are also missing in members of other taxa that inhabit lightless, or very low-light, environments. For example, there are blind lobsters in deep seas, blind cave crickets, subterranean ants that lack the usual insect compound eyes, blind moles (along with partially sighted ones), eyeless cave-dwelling spiders, and some blind velvet worms that live under stones and in other low-light environments.

Even in those cases in which blindness is accompanied by complete eye loss, and so there is no developmental job to do in terms of making an eye, the *Pax6* gene plays multiple other roles, including various roles in the nervous system, plus several other non-neural roles. The latter include, in mammals, roles in the gut and the pancreas. Studies on mice have shown that *Pax6* is co-expressed with insulin-producing genes. This illustrates a link-up between transcription factors and hormones (insulin is a peptide hormone). Although I have not mentioned hormones up to now, these too are important players in developmental processes, and not just in vertebrates. In insects, the interplay between juvenile hormone and the moulting hormone ecdysone is instrumental in determining the series of moults that occurs in post-embryonic development.

Like Hox genes and most other toolkit genes, *Pax6* has many downstream targets. This is even true when we are considering just one developmental role – in eye development. Some of the genes that *Pax6* switches on are themselves transcription-factor genes; others are genes making eye-specific proteins such as the crystallins, which make up much of the lens and cornea of the vertebrate eye. In both eye development and neural development it seems that *Pax6* switches on various genes that make cell–cell adhesion proteins. Like hormones, these are also key players in development that I have not mentioned up to now. One of the ways in which blocks of tissue stick together is by cells of a particular type all being characterized by the same type of adhesion molecule on their outer membrane, while other cell types are characterized by different adhesion molecules – or different combinations of them.

To conclude, *Pax6/eyeless* is a key gene in eye development and in several other developmental processes. It is probably found in all bilaterian animals. Eyes are not found in all of these animals, but some sort of eye is present (in various numbers) in most of them. The question of whether all eyes are homologous is a difficult one. Perhaps all or most animal eyes share at least a common ancestral rudimentary light-sensitive device. However, there are some instances in which a complex eye type has clearly arisen convergently from a simple one – most obviously the camera-type eyes of vertebrates and cephalopods, as noted above.

Gene Co-option in Evolution

As we have seen, there are many cases in which it seems that a toolkit gene that started off with one developmental function at one point in evolution later acquired others. This process of genes being co-opted, or recruited, for new functions is thus an important mechanism involved in the evolution of development. So we need to understand how it works. In terms of the molecular details, it may work in various ways, but what we need is a general outline of the process, which is what follows. For more details, see the reviews by John True and Sean Carroll (2002), and Chris Jiggins and colleagues (2017).

The evolutionary co-option of a gene for a new developmental role typically involves mutational change in a regulatory rather than a coding region – a change that results in the gene being switched on at times, or in places, when/where it was previously switched off. We have seen this distinction between regulatory and coding regions before, but now we need to look at regulatory regions in more detail. Every gene has a regulatory region called a promoter, which is located at one end of the coding region (the end at which transcription will start). When the gene is switched on, an enzyme called RNA polymerase attaches at this site and moves along the coding region, transcribing it into RNA. The switching on is governed by the arrival of transcription factors, as we have already seen. Some transcription factors bind to the promoter region; however, many of them bind to other regulatory regions called enhancers. While an animal or plant gene often has just one promoter, it typically has several enhancers. In the early days of molecular genetics, it was thought that these resided just 'upstream' of the promoter (in terms of the direction of transcription). However, it is now clear that they can exist both there and at other sites within the gene, or even elsewhere on the chromosome. Looping of the DNA can bring enhancers that are not physical neighbours into temporary contact.

Not only are there multiple enhancers, but each has its own characteristics. In particular, one of them might have a sequence such that 'transcription factor X' binds to it (Figure 8.1). Imagine that the gene concerned initially has a particular role in development in the embryonic head, and that in this part of the embryo the gene for transcription factor X is expressed. If attachment of this particular transcription factor is sufficient to switch the target gene on (usually more factors are required, but recall that this is a simplified general outline only), then it will indeed be switched on in the embryonic head and hence it will be able to fulfil its developmental role.

Now suppose that, sometime later in the evolution of the lineage concerned, one of the enhancers duplicates. Recall from Chapter 6 the importance of the duplication and divergence of genes in evolution. Duplication and divergence of parts of genes, and in particular their enhancer regions, is also important in evolution – and for the same reason. Once an enhancer has duplicated, there is relaxed selection on one of the duplicate copies and it is likely to accumulate more mutations than usual because these are not removed by 'purifying' selection. Often, this will simply lead to the enhancer becoming dysfunctional. But sometimes its new sequence will be such that another transcription factor (Y) can bind to it (Figure 8.1). The gene will then not only be expressed in the embryonic head where X is available, but also in other regions of the embryo (say the flanks of the embryo) where Y is available (and providing that Y is able to switch the gene on by itself, which again is a simplification). We can see from this hypothetical example how an evolutionary change in the regulatory regions of a gene can enable it to be co-opted for a new role in a part of the embryo where the gene used to be silent.

So far, the story has involved just the gene we're focusing on and a couple of genes that are upstream of it in the developmental cascade/hierarchy – the ones that make the transcription factors that switch it on. However, the other side of the story – the downstream side – is equally important. If our gene-in-focus becomes switched on in parts of the embryo in which it was previously (in an evolutionary sense) switched off, what will be the developmental result of this? On the basis of the information provided so far, this question cannot be answered. To answer it, we need to know what our gene does. For example, it might be a gene that is involved in the directionality of cell

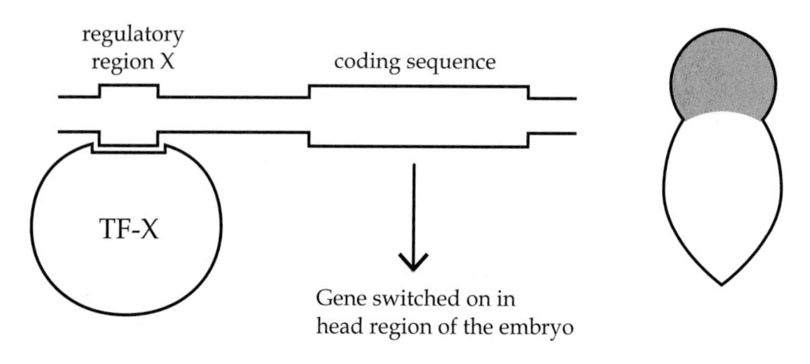

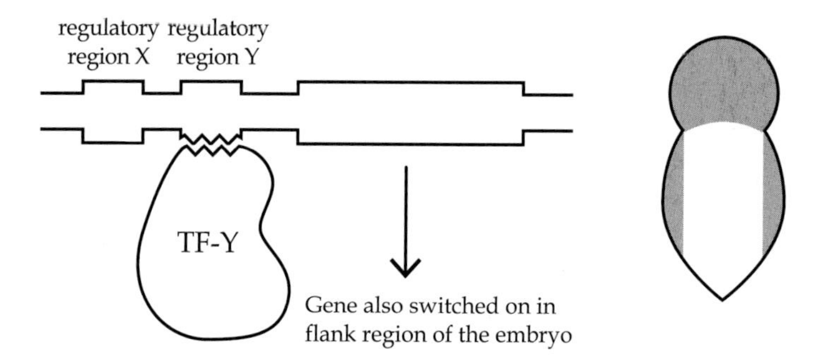

Figure 8.1 A way in which gene co-option might occur in evolution. Top: evolutionary starting point at the level of the gene, showing an ancestral regulatory region, or enhancer, X. The transcription factor (TF-X) that binds to this enhancer is present in the head of the embryo (right), so the gene is switched on there. Bottom: evolutionarily derived version of the gene, with a second regulatory region, or enhancer, Y, that has arisen from duplication and divergence of the original one. A different transcription factor (TF-Y) that binds to this enhancer is present in the flanks of the embryo, so the gene is now switched on there too.

proliferation. Its new role might involve causing cells to grow out from the sides of the body, producing limbs, while its old role perhaps helped nerve cells to grow in the right directions to form neural circuits. Although this is a hypothetical example, something rather like this may have happened in the case of distal-less genes. If *distal-less* was first expressed in the nervous system and then later co-opted for a new role in the limb buds, the above scenario could apply. But the gene-in-focus could be any gene at all – the idea of gene co-option is a very general one.

The co-option scenario described above is an example of heterotopy – the form of developmental repatterning that involves changes in space. However, gene co-option is equally involved in heterochrony – changes in time. The logic of the argument is the same. The only difference we need to invoke for heterochrony to be the result is that transcription factor Y is available (i.e. its gene is switched on) at a different point in developmental *time* than X, rather than them being available in different *parts* of the embryo. And the other two types of developmental repatterning can easily be brought into the picture as well. Evolutionary changes in regulatory regions of a gene can alter its degree of affinity for a particular transcription factor rather than switch its affinity to another one – hence we would have heterometry. And one type of transcription factor can itself evolve into another type by evolution of the coding regions of the gene that produces it – heterotypy.

An important difference between the last of these kinds of evolutionary change and the other three is that it is less flexible. Because an evolutionarily altered transcription factor has a different sequence of amino acids, it may bind to different regulatory sequences of other genes – but it will potentially do this at all developmental stages and at all developmental times. This is why many evo-devo authors, notably Sean Carroll, have emphasized the 'primacy' of changes in regulatory rather than coding regions of genes.

The whole suite of toolkit genes that we recognize in today's animal kingdom has itself arisen by duplication and divergence of a small number of toolkit genes that existed in the very first animals. Over the long course of animal evolution – some 600 million years or so, as we have seen – the toolkit has been enlarged in its number of tools, and many of the tools have adopted more roles via gene co-option. Many of the roles that developmental genes have in complex animals such as today's vertebrates and arthropods did not exist in our distant bilaterian ancestors. So roles themselves originate and evolve. Or, to put it another way, the genome, developmental processes, the overall developmental trajectories of which these processes are a part, and the wonderful forms of animals that we see around us – both in developmental stages such as caterpillars and in adult forms such as butterflies – all evolve together.

Finally, since the phrase 'toolkit genes' is now firmly established, there is one important semantic point to note: genes are not really tools. Tools are things that are employed to physically get a job done. Thus in the body, proteins can be thought of as molecular tools. A homeodomain protein is a tool used to switch downstream genes on. The toolkit genes are the codes from which the tools get made. Or more precisely their coding regions are the codes from which the tools are made. Their regulatory regions are harder to classify in this respect. Is a transcription factor a tool, or is a regulatory region of a target gene for that factor a tool, in that it grabs the factor as it is passing by? Ultimately, it is neither codes nor tools on their own that are the key players in development and its evolution, but the interactions between them.

9 Bringing It All Together

A Combined View of Development

Taking an evo-devo approach to the living world changes our views of both of biology's great creative processes – evolution and development. However, this is not a symmetrical change. Evolutionary biology is a more theory-driven discipline than developmental biology. Why this is so is itself an interesting question, but perhaps more one for philosophers and historians of biology than for biologists themselves. Anyhow, this difference between the disciplines means that our altered view of development is easier to adopt than our altered view of evolution – in the sense that it causes less tension in a pre-existing theoretical framework. Because of this fact, I will start with the simpler case (development, this section) but will spend longer on the more complex one (evolution, the other sections).

Before the advent of evo-devo, developmental biology texts described the development of a relatively small number of organisms in what might be called splendid isolation. There was little or no phylogenetic context. We learned from these texts, for example, how 'the frog' developed, based largely on work conducted on the model system *Xenopus laevis*. There was generally little attention paid to the fact that 'frog' refers not to a single species but rather to a clade of several thousand species – estimates vary, ranging up to about 7000. Nor was there much attention paid to the fact that within this large clade there was considerable variation in developmental trajectories. We looked at one of the most notable aspects of this in Chapter 1 – the fact that some frog species have evolved from indirect to direct development – they have lost their tadpoles, evolutionarily speaking.

The situation was much the same when other model organisms were discussed in textbooks. For example, some aspects of chick development are unique to their own species, subspecies *Gallus gallus domesticus*, and even strain, while some are shared more widely, right across the avian clade (about 10,000 species) and often beyond. A case in point is Sonic hedgehog signalling from the notochord, which we looked at in Chapter 3. This signalling is similar, at least in its broad mode of operation, across different vertebrate model systems, including the chick, the frog, and others. The problem is that when we consider the development of any one kind of animal in isolation from others we don't know the degree of breadth of applicability of *any* aspect of its development.

This is a good point at which to recall the motivation of the experimental embryologists of the early twentieth century. They were motivated by a desire to understand the causal mechanisms of development; this was very sensible, and allowed them to make major advances such as the discovery of the 'organizer', which we noted in Chapter 2. To make such discoveries, it was necessary to leave behind the comparative embryology of von Baer, Haeckel, and others, and to concentrate on model systems. But, now that we have lots of causal explanations for developmental processes in such systems, it makes sense to bring phylogeny back into the picture. That way, we end up with a combined view of development: an understanding of the causal mechanisms at work in individual species, *but also* an understanding of the way in which some of those causal mechanisms apply to quite restricted clades, others to much broader ones. This combined information is instructive about both development *and* evolution.

Towards a More Comprehensive Evolutionary Synthesis

I have mentioned in passing a couple of times in earlier chapters the twentieth-century 'modern synthesis' of evolutionary theory. Now we need to look at its origins. While the early experimental embryologists were beginning to discover causal mechanisms of embryogenesis, the early population geneticists were beginning to put together a synthesis of Mendelian genetics and Darwinian natural selection. The first key book of this synthesis was *The Genetical Theory of Natural Selection*, by Ronald Fisher, published in 1930, which was closely followed by *The Causes of Evolution*, by J. B. S. Haldane, published in 1932. Also key in the population genetics synthesis was the work of Sewall Wright; a major paper of his called 'Evolution in Mendelian populations' appeared in the journal *Genetics* in 1931. These three scientists are often referred to as 'the great triumvirate' of population genetics.

A genetically based theory of natural selection has an advantage over a merely descriptive theory because it is quantitative. This is just one example of a general principle in science: a quantitative theory, where one is possible, is usually preferable to its qualitative counterpart; it makes more precise predictions and hence is more readily testable. However, many evo-devo biologists feel a greater affinity with Charles Darwin than they do with theoretical population geneticists, whether the pioneering ones of the 1930s or their academic progeny of subsequent decades. Why is this?

In my view, it's because although something was gained in the transition from a Darwinian approach to a population genetics approach, something was also lost: the organism. Our quantitative theory of evolution was formulated largely in terms of two levels of organization – the gene and the population. The level in between – the developing organism – was neglected, at least in comparative terms. As the 1930s theoretical population genetics synthesis (of Mendelism and Darwinism) gave way to the broader 'modern synthesis' of the 1940s and 1950s, this problem was reduced, though not enough. The broader synthesis incorporated field-based population genetics (Theodosius Dobzhansky), systematics (Ernst Mayr), zoology (Julian Huxley), botany (Ledyard Stebbins), and palaeontology (G. G. Simpson) – but developmental biology was left out in the cold. Or was it?

The answer to this key question is: 'not quite'. Julian Huxley's 1942 book *Evolution: The Modern Synthesis* is interesting in this respect. In his preface, Huxley says:

> Any originality which this book may possess lies partly in . . . stressing the fact that a study of the effects of genes during development is as essential for an understanding of evolution as are the study of mutation and that of selection.

The thought was there, but it wasn't really followed by action. There's a short section on development in Huxley's Chapter 9 ('Evolutionary trends'), but not much else.

One of the embryologists to whom Huxley refers in his book is Gavin de Beer. We looked briefly at de Beer's work in Chapter 2, especially *Embryology and Evolution*, published in 1930 and republished in 1940 as *Embryos and Ancestors*. Huxley refers to both of these versions. Why is de Beer's book not usually mentioned as a major contribution to the modern synthesis? Well, partly because he himself made little attempt to connect his own approach with the synthesis. Even when the final edition of *Embryos and Ancestors* appeared in 1958, there was almost no connection made. For example, the only work of Fisher's that de Beer refers to is a paper on 'Dominance in poultry'.

So, while developmental biology was not completely left out in the cold by the evolutionary synthesis of the 1930s to 1950s, it was not warmly embraced either. Ever since its quasi-exclusion, various biologists have argued that the synthesis should be enlarged to bring it into the fold. We've already looked at some of these biologists, including Conrad Hal Waddington and Stephen Jay Gould. Their respective books *The Strategy of the Genes* (1957) and *Ontogeny and Phylogeny* (1977) influenced the thinking of many biologists. But the calls for a more comprehensive synthesis really took hold with the advent of evo-devo in the 1980s. From then until now, there have been many such calls; however, the people making them are not all singing from the same hymn sheet. This is reflected in two things – the names they give the sought-for new synthesis and the exact developmental (and other) themes they wish to see given pride of place.

The palaeontologist Robert Carroll wrote a paper in 2000 called 'Towards a new evolutionary synthesis'; his focus was on both developmental biology and palaeontology. The evo-devo proponent Sean Carroll (no relation!) wrote a paper in 2008 entitled 'Evo-devo and an expanding evolutionary synthesis'; his focus was very much on molecular developmental genetics, and in particular on the importance of evolutionary changes in the regulatory rather than coding sequences of toolkit genes. The theoretical biologists Massimo Pigliucci and Gerd Müller co-edited a 2010 book called *Evolution: The Extended Synthesis*. This was more eclectic in its messages about a more comprehensive synthesis than either of the Carrolls' papers were; but this is hardly surprising, given that it encompassed the work of many authors, including philosophers as well as biologists.

So, what I call a 'more comprehensive' synthesis has been referred to by others as 'new', 'expanding', 'expanded', and 'extended'. In my opinion, these terms are all equally good. But they have not been equally effective, as is easily seen by doing a quick Google search. By that measure, 'extended synthesis' wins. There is even a Wikipedia entry for 'extended evolutionary synthesis', while there is not (at the time of writing) for the others. And it has generated opposition as well as support. There is a strange double article in *Nature* in 2014, one that falls under the unusual heading (for that journal) of 'Comment'. The title is 'Does evolutionary theory need a rethink?' One part of the double article, whose lead author is Kevin Laland, says 'Yes, urgently'. The other, whose lead authors are Greg Wray and Hopi Hoekstra, says 'No, all is well'. Unsurprisingly, one of the co-editors of the *Extended Synthesis* book is on the 'yes, urgently' team.

In my view, the main thing wrong with 'the extended synthesis' is its use of 'the'. It is preceded by the definite article in the title of the 2010 book, on its Wikipedia page, and in many published articles. Use of 'the' could be taken to imply that an extended synthesis exists, is clearly formulated, and is unopposed, but arguably none of these things is correct. So, although I have no preference between, for example, 'expanded' versus 'extended', I most certainly do have a preference for the indefinite article. The Carrolls' use of 'a/an' seems more appropriate, at the current stage of play, than Pigliucci and Müller's 'the'. In the rest of this chapter, I will try to summarize and draw together potential evo-devo contributions to *a* more comprehensive evolutionary synthesis.

Variation and its Interaction with Selection

In Darwin's *Origin of Species*, the chapter on natural selection is preceded by two chapters on variation and one on the struggle for existence. Immediately after the natural selection chapter, he returns to variation in one entitled (unwisely perhaps) 'Laws of Variation'. So no-one could say that Darwin downplayed variation compared to selection. Yet as the 'modern synthesis' took shape, and Darwinism gave way to neo-Darwinism, such downplaying was exactly what happened. This is not to say that the architects of the modern synthesis *wanted* to downplay variation – they didn't. Rather, they regarded many aspects of it as being inaccessible to productive study at the time. We noted this in Chapter 5, in relation to J. B. S. Haldane's comment that 'only a thorough-going study of variation will lighten our darkness', but that biologists of his day could only 'dimly conjecture' about this important matter. In the era of evo-devo, we can do much better than dimly conjecture. We can get to grips with the nature of variation, and we can do so at different levels, both molecular and morphological. It is, of course, at the molecular level that our knowledge has increased the most since Haldane's time.

What we now understand – but did not prior to the 1980s – is that a loosely defined group of genes, the toolkit genes that I have discussed in several of the previous chapters, are involved in development across most or all of the animal kingdom, and in some cases beyond. These genes are thus described as being characterized by the deep homology that we encountered in Chapter 8. Ancestral genes were present in an early stem lineage, and their descendant genes among today's fauna have arisen from those ancestral genes by various processes, notably mutation, gene duplication, and higher levels of duplication all the way up to the genome.

Mutations in toolbox genes can have major effects on development, and hence on the adult phenotype – if indeed they permit development to proceed that far. Major mutational effects on development typically go hand-in-hand with severe reductions in fitness. Such mutations have been of much use in revealing the developmental roles of the normal versions of the genes concerned, but they are by definition not involved in adaptive evolutionary change.

However, this does not mean that *the genes themselves* are irrelevant to evolution. Indeed, their involvement is logically necessary – well, depending on how toolkit genes are defined. If we regard 'toolkit gene' and 'developmental gene' as being synonymous, then the evolution of these genes necessarily goes hand-in-hand with the evolution of the development (i.e. the four-dimensional form) of the animal.

Perhaps the biggest challenge at the molecular level in the 2020s is to acquire a thorough knowledge of how toolkit genes vary within and between natural populations, and of how this molecular 'standing variation' links with variation in developmental pathways. This endeavour needs to extend widely from a phylogenetic perspective – that is, there must be appropriate taxon sampling. We need to know not just what the situation is in *Drosophila* and other model systems, but across the board. This is especially important given that, as we have seen, model systems are atypical in many ways.

Although 'developmental bias' normally refers to variation at the phenotypic rather than genotypic level of organization, as in the cases of mammalian neck vertebrae and centipede trunk segments, bias *sensu lato* is also applicable to variation at the level of individual genes. This is sometimes called 'mutation bias', but care is needed with this term as it has been used in different ways by different authors. The version most relevant here is that used in 2001 by the American theoretical biologist Arlin Stoltzfus, who argued, with his colleague Lev Yampolsky, that while some evolution is based on the 'standing variation' mentioned above, some is based on newly occurring mutations, and in this case the order in which mutations occur can be important.

Sean Carroll's emphasis on the importance of evolutionary changes in the regulatory as opposed to coding regions of developmental genes might at first sight seem like a mutation bias, but that's certainly not what he intended to imply. Rather, it is a selective bias. In other words, instead of mutations occurring preferentially in those regions, mutations occur equally there and in coding regions, but while selective constraint often prevents their spread in the population when they affect the structure of the protein, such constraint may be weaker when they don't. The rationale here is that a mutant protein will be mutant everywhere in the body, with likely deleterious effects, while a non-mutant protein produced by a molecular interaction happening at a mutant regulatory region may have its quantity, timing, and location altered but not its fundamental nature. This corresponds to greater ease of evolution by heterometry, heterochrony, or heterotopy than by heterotypy.

Note that the distinction between mutation bias and selection at the level of a gene is clear-cut. But recall that the higher-level phenomenon of developmental bias/constraint is not clearly distinct from selective bias/constraint. This difference arises from the fact that whereas a gene is made quasi-instantaneously, an organism is made through a long process of development. As we saw in Chapter 4, what looks like developmental constraint from the perspective of one stage of development can look like selective constraint from the perspective of another.

Variation at the molecular level is an important piece of the whole jigsaw of evolutionary change. Yet normally the link between such variation and

Darwinian selection is the phenotype – or the animal's development, to think in four-dimensional terms. Arguably the most important thing evo-devo has done from a general philosophical perspective is to put the animal (or plant) back at the centre of our attention. Accordingly, we now move on to consider variation at the level of developmental trajectories; this takes us back to the concept of developmental repatterning.

Developmental Repatterning

As noted in Chapter 4, it is important to have a cover-term for changes in development – repatterning – just as it is important to have a cover-term for changes in a gene – mutation. Having such terms enables a balanced approach to evolutionary change in which neither the gene nor the organism is downplayed. However, we should pause at this point to reflect on the importance of inheritance in both cases. Even gene mutation is not necessarily inherited. When we speak of the role of mutation in evolutionary processes, we are implicitly focusing on mutations that occur in cells of the germ line, not in cells that lead only to a group of somatic cells ('a clone') that make up some part of an organism's body (the soma) and die along with the rest of the organism when the time comes. As an aside here, it is worth emphasizing that in some cases it is the mutant clone that actually *causes* the organism's death – for example when a somatic mutation leads to an aggressive tumour.

Like mutation, developmental repatterning is not necessarily inherited. It is what it says it is – a repatterning (of any kind) of part or all of the developmental trajectory from zygote to adult. This can occur for genetic or environmental reasons. The repatterning of fruit-fly development that leads from the possession of normal functional wings to vestigial functionless ones is certainly inherited – though this particular repatterning is probably not relevant to evolution because the mutation is detrimental to fitness, even in environments in which flying is unnecessary. In contrast, the repatterning of fruit-fly development caused by severe larval food limitation that leads to smaller adult body size is not inherited. Smaller flies lay fewer eggs, but if the larva of such an egg has plentiful food, the developmental repatterning will be reversed. Thus a change that occurs in one generation may have no consequences for the next – unlike the situation with mutationally based repatterning.

There are two complications that we need to take on board at this stage. The first involves developmental reaction norms, discussed in Chapter 5. In discussing these, we saw that there can be an *interaction* between genetic and environmental causation in producing developmental repatterning. A particular developmental trajectory is the result of possession of a particular genome and growing up in a particular environment under a particular set of conditions. Change either the genome or the environment, and repatterning may take place. Although consideration of such interactions has broadened our view of the role of the environment in evolutionary change, starting with the experiments of Waddington, it is important to realize that if there is no genetic variation in the first place there will be no possibility of a pattern of phenotypic plasticity to evolve. This takes us back to the simple but key point – inheritance is essential if evolution is to occur.

The second complication to the concept of developmental repatterning is something I haven't stressed up to now, though it was lurking in the background when I mentioned the work of Arlin Stoltzfus in the previous section. If a change in development is called repatterning, what was the starting point – the 'original patterning', if you like? In broad comparisons, the answer to this question is clear, while in narrower, more tightly focused comparisons it is not. One example of the former would be a comparison between an ancient ancestor of manatees, with seven neck vertebrae, and its descendants, with six. Another example would be the dextral ancestor of a clade of gastropod species that are all sinistral. This second example can even be seen on a narrower basis, for example between the 'normal' dextral form of the water-snail *Radix peregra* and the sinistral variant that is only found in a very small number of populations.

However, what if we are considering two very similar but non-identical developmental trajectories that characterize two individuals within a particular natural population of a particular species? It doesn't matter whether the species is a manatee, a snail, a beetle, or an elephant. In this case, there is no identifiable original trajectory – rather, each can only be considered arbitrarily as a repatterning of the other. Expanding our view from 2 to *n* individuals in the population, where *n* could be 3 or 1000, any individual trajectory could be regarded as a repatterning from the perspective of any other. Perhaps this is straining the term repatterning too far, but then again as long as we realize that everything is relative when we cannot identify a particular *original* trajectory, perhaps that's fine.

Finally, it's worth recalling that developmental repatterning is, like mutation, intended only to be a descriptive cover-term. It is, as I have called it earlier, theory-neutral. Its use is important to give the organism and its development parity of treatment with the gene. But using the term does not imply any particular driver of evolutionary processes. When population geneticists refer to mutations, they are usually not advocating evolutionary mutationism (i.e. saltationism). When proponents of evo-devo refer to developmental repatterning, they are not *necessarily* advocating a 'constraintist' approach to evolution in the sense of biased production of developmental repatterning steering evolution in particular directions. However, they might be; let's see.

Developmental Bias and the Direction of Evolution

The question of what 'steers' evolution in one direction rather than another is especially important. When Darwin described natural selection as the 'main' mechanism bringing about evolutionary modification, he meant, primarily, that it is steering the process. It may be doing other aspects of driving as well, for example sometimes accelerating and sometimes braking, but it rarely takes its hands off the steering wheel. Having said that, Darwin is often described as a 'pluralist': he did not see selection as the *sole* driver of evolution. There are many aspects of this issue, including the question of whether he was too ready to acknowledge a role for 'use and disuse' (the inheritance of acquired characters) in later editions of *The Origin of Species*. But here we will focus on just one aspect of Darwin's stance, namely his emphasis on the importance of what he called 'correlation of growth' (which we first encountered in Chapter 5).

What Darwin meant by correlation of growth was that development is a complex process, and that if one aspect of it becomes altered, for example by selection, then other aspects are likely to change too. In other words, each measurable character is necessarily correlated with many others. Here's what he says at the start of the relevant section:

> the whole organisation is so tied together during its growth and development, that when slight variations in any one part occur, and are accumulated through natural selection, other parts become modified. This is a very important subject, most imperfectly understood.

It is here, perhaps, that the views of Darwin and Wallace differ most dramatically (apart from Wallace's exclusion of the human mind from the domain of natural selection). In contrast to Darwin's emphasis on correlation of growth, Wallace emphasized the independent variation of different characters.

One aim of evo-devo is to probe Darwin's 'correlation of growth' in development and to assess its role in the evolutionary process. However, a difficulty in doing this is the existence of many related terms with meanings that overlap with each other to a variable degree, depending on the author. Here they are (and this is by no means an exhaustive list):

- correlation of growth
- correlated characters
- developmental correlation
- developmental constraint
- developmental bias
- developmental drive
- developmental channelling

Given this plethora of terms, let's for the moment see how far we can get without using any of them except for Darwin's 'correlation of growth'. The best way forward is to think of a spectrum of possible modes in which evolution might work, characterized by two extremes, as follows.

1. At one end of the spectrum, selection can modify development in any way it likes. Either there is no correlation of growth or, if there is some such correlation, there is also sufficient variation in the pattern of combined character values among different individuals for selection to be able to reduce the correlation to zero, thus rendering any character potentially modifiable on its own, without the rest of development being affected.
2. At the other end of the spectrum, the development of an animal is so tightly integrated, and various developmental characters are so correlated and interwoven, that selection cannot make independent changes in any character without necessarily modifying many others. Selection's scope is therefore very limited. In this case, the few directions in which selection can proceed are predetermined by the nature of the developmental process, so much so that the role of 'steering' evolution lies more with development than with selection. You will recognize this latter

extreme view as the one adopted by the American biologists Stephen Jay Gould and Richard Lewontin (discussed in Chapter 4).

The truth is almost certainly in between these two extremes. That's the easy bit. Determining exactly where it lies between the extremes, and what exactly are the dynamics of the interaction between development and selection, is the hard bit.

The butterfly experiments conducted by Paul Brakefield, Particia Beldade, and their colleagues that we discussed in Chapter 5 were particularly informative in this respect. As we noted there, these experiments gave two contrasting results: on the one hand, selection was able to break the correlation between the sizes of different eyespots on the butterflies' wings; on the other hand, it was not able to break a correlation between the pigmentation patterns of different eyespots. If this result is applicable beyond the characters and species studied, it confirms the 'intermediate' nature of the relationship between the dynamics of development and the dynamics of natural selection in determining the way in which evolution proceeds. Reality does not reside at either end of the spectrum but somewhere in between them.

Those butterfly experiments were pioneering studies of the way in which correlation of growth and natural selection interact with each other, and they were very informative. However, many more studies are needed. Ideally we would like to know about thousands of characters in thousands of species rather than just a few characters in one species, though the work involved would be prohibitive. Also, multi-generation selection experiments are feasible with some species of insects, but in many other animal groups – primates, bats, and deep-sea tube-worms, to name but three – they are most certainly not. So we need to be realistic. It is within the realm of realism that many more characters can be investigated in this way in many carefully selected species. When such investigations have been done, we may have a clearer view of where on the spectrum between the two extremes of pan-selectionism and 'pan-constraintism' (as we can call the other end of the spectrum) the interaction between developmental bias and natural selection falls.

The Origins of Novelties and Body Plans

A key question that arises at this point is whether the interaction between the biased supply of developmental variants and their biased survival or reproduction (selection) is somehow different when we are considering the (occasional) origins of novelties or body plans, instead of considering what can be called routine evolution – the sort of evolutionary change that is going on in most lineages most of the time. I do not think we are in a position yet to give a satisfactory answer to this question. However, some tentative comments can be made, as follows.

First, the origin of a novelty, such as the shell of turtles (Chapter 6), or a body-plan feature, such as segmentation (Chapters 7 and 8), is something that strikes most practitioners of evo-devo as being of particular interest. Such an origin seems to be indicative of the occurrence of processes over and above those that are involved in, say, the evolution of slight modifications in body size or shape. However, this is a subjective view, so we should treat it with caution.

Second, if indeed there is something different about these interesting evolutionary origins, it is more likely to lie in the realm of development, and the availability of variation in development, rather than in the realm of selection. This possibility is illustrated in Figure 9.1, which shows a comparison of the availability through evolutionary time of the requisite variation to allow the evolution of a larger beak size in birds on the one hand and the origin of a turtle shell on the other. The former variation is omnipresent, while the latter occurs only in fleeting bursts. It is this distinction that causes the rarity of inventions of an integral shell (only once in vertebrates, albeit in stages, with the plastron first), in contrast to the commonness of trends towards larger beak size (too numerous to count). Of course, the situation pictured here is only a hypothesis; nevertheless, it seems more plausible than the alternative one, that selection for a protective shell is very rare.

Third, although the availability of unusual kinds of variation may be the area in which we should look for explanations of the origins of novelties and body plans, there is no simple distinction such as small-effect versus large-effect mutational changes in development. Back in the middle of the twentieth century Richard Goldschmidt (Chapter 2) claimed that there were two radically different kinds of evolutionary change – micromutational change producing the evolution of races within a species, and macromutational change producing more significant evolutionary origins. This saltational theory of evolutionary origins was not acceptable to most biologists in Goldschmidt's

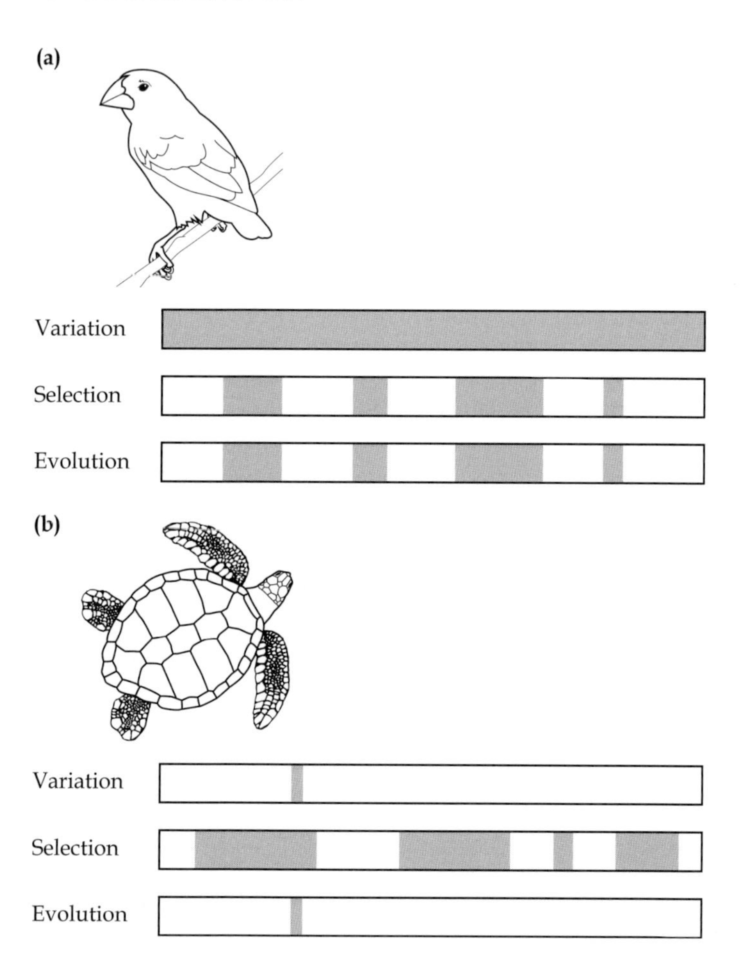

Figure 9.1 Hypothesis of the contrasting ways in which (a) routine evolution and (b) the evolutionary origins of novelties (or body plans) might work. In the case of routine evolution, such as a bird lineage evolving a bigger beak (top), the variation is omnipresent, so the periods of time during which evolution produces larger beak size correspond with periods of selection for larger beaks. In the case of the origin of a novelty, such as the evolution of a shell in the turtle stem lineage, the necessary variation is rarely present, so the period during which the shell originates corresponds with the period of available variation (selection for a protective shell probably being common through evolutionary time, as indicated). These two scenarios could be called 'selection-limited' and 'variation-limited' evolutionary change respectively.

time, and no subsequent discovery has made it any more acceptable to the majority of biologists today. As we have seen, there are a few exceptions in which a 'macromutation' *sensu lato* is involved in evolution, such as the one that transforms a sinistral gastropod into a dextral one or vice versa. The apparent doubling of the number of trunk segments in the origin of the centipede species *Scolopendropsis duplicata* is probably another example. But macromutations *sensu stricto*, involving more radical change in body form than a symmetry reversal or a change in segment number, may not have been involved in evolution at all. So far, there is no evidence that they have been.

Finally, we really need to get away from the simplistic notion of micro- versus macromutation. Associated with this, we need to get away from the notion of a *single spectrum* of magnitudes of developmental effect of mutations that has micro- and macro- at its opposite ends. We need to think instead about the *type* and *timing* of mutational effects on development. Then all sorts of interesting questions begin to emerge. For example:

- Do the origins of evolutionary novelties and body plans typically involve variation in earlier developmental stages than the variation that acts as a basis for routine evolution in body shape?
- Is the balance between the types of developmental repatterning that we call heterochrony, heterotopy, heterometry, and heterotypy different when we compare 'origins' with routine changes?
- At the molecular level, might there be a difference in the importance of gene co-option between the two situations?

As evo-devo progresses we may eventually be able to answer these and related questions. This is an exciting prospect, but we shouldn't underestimate the magnitude of the challenge. The important thing to note at this stage is that these are all questions about probabilities and balances, not about stark black-and-white contrasts.

The Pattern of Occupation of Morphospace

Looking across the animal kingdom, we can easily see that the forms of actual animals represent many variations on relatively few themes. Related to that, there is an absence of many conceivable animals (e.g. six-legged mammals), despite their apparent potential functionality from an engineering point of view. One way of describing this kind of pattern is to say that the actual forms of animals that evolution produces are very non-randomly distributed across all possible morphospace. The importance of this point was emphasized in 2002 by Stephen Jay Gould, in his magnum opus *The Structure of Evolutionary Theory*. He put it like this (page 1174):

> The markedly inhomogeneous occupation of morphospace – surely one of the cardinal, and most viscerally fascinating aspects of life's history on earth – must be explained in terms of historical constraint, and not by the traditional mapping of organisms upon the clumped and non-random distribution of adaptive peaks in our current ecological landscapes.

Gould's use of 'viscerally fascinating' strikes a chord with me. I can't think of a better way to put it. And while gut instincts are not always to be trusted, their importance to the real scientific process (as opposed to the cartoon version) should not be underestimated. However, I disagree with Gould's statement about how the non-random pattern of occupation of morphospace might be (or 'must be', in his words) explained.

Gould sets up a pair of apparent opposites: historical constraint on the one hand and selection acting under present-day ecological conditions on the other. But this is not a sensible contrast, because it's hard to see how anyone could believe that selection acting in the present could produce a pattern that has clearly built up over the long-term past. Logically, then, the cause must have a historical dimension, whether it is based on selection or constraint. With this modification, then, our two contenders to explain the overall pattern become historically accumulated selection versus historically accumulated constraint. Another twist in identifying the contenders is that the relationship between them may not be either–or (versus) but both–and (interaction between the two). And a final twist is as follows. We've seen that what is often called developmental constraint is better described as developmental bias, incorporating both positive effects (developmental drive) and negative ones (constraint *sensu stricto*); we've also seen that selection can be positive (directional) or negative (stabilizing). So the interaction is not between a negative force and a positive one, but rather between the non-random production of variants (bias) and their non-random survival and reproduction (selection). Both of these can act in a positive or a negative way. The key issue is to understand the dynamics of their interaction.

Now is a good time for us to remind ourselves that an animal form is actually *not* a single point in morphospace. This might be true if we take a three-dimensional, adultocentric view. But taking a four-dimensional view, an individual animal is itself a highly non-random *developmental trajectory* through morphospace. And although this trajectory has been moulded by the accumulated natural selection and developmental bias of the past, its own particular arrow through time and space is driven by neither of those processes. Rather, it is driven by the interactions among genes, proteins, cells, and other players in what we might call the developmental dance. And included among the 'other players' is the environment – recall the interaction between environmental and genetic variables that we saw when discussing phenotypic plasticity in Chapter 5.

The magnitude of the task facing practitioners of evo-devo should now finally be clear. We seek to understand the interaction between two sets of interactions: those that steer the developmental trajectory of an individual animal or plant in the short term, and those that steer the repatterning of such trajectories in the context of natural populations of animals and plants in the long term. As intellectual challenges go, they don't come much bigger. Our efforts necessarily span many levels of biological organization as well as many timescales. Our job, as in other areas of science, is to uncover the *general principles* that govern our 'interaction between sets of interactions', rather than to make a long list of details. However, our quest for generality should go no further than the real world permits. Evolution and development are both probabilistic processes; so we should not seek universal laws. We should learn from the lessons of the past – in particular the supposed 'laws' of comparative embryology. There is no evo-devo equivalent of $E = mc^2$. But that should not concern us. General principles can possess beauty too, despite the occurrence of things that we might consider to be ugly exceptions. In the living world, general principles are more likely than proposed 'laws' to capture the messy thing that is reality. Science is a search for both truth and beauty, but in cases of conflict we must give priority to the truth.

Concluding Remarks

Evo-devo has come a long way since its origins a mere four decades ago. Many exciting things have been discovered, and there will be many more discoveries to come in the years ahead. I have tried, in this book, to give you a flavour of this new branch of science. Here, I summarize what I think are its most important conclusions so far and the most important challenges that lie ahead.

From a literary perspective, it is sometimes said that a book arriving back where it started has an aesthetic appeal. I can easily oblige the aestheticists in this case because I mentioned in the Preface that one of the key discoveries of evo-devo – some would say *the* key discovery to date – was that very different animals have very similar developmental genes. This was completely unexpected. Animal forms range from millimetre-long worms to mega-size blue whales, from soil centipedes to soaring eagles, from polar bears to the various denizens of the super-hot hydrothermal vents at the bottom of the oceans. Many biologists had assumed, prior to the advent of evo-devo, that the development of such different forms would require very different developmental genes. Instead, it turns out that members of a rather small number of families of toolkit genes are found throughout the animal kingdom, governing the dynamics of the developmental trajectories that produce such radically different creatures. This applies to the three main body axes of bilaterally symmetrical animals – anteroposterior, dorsoventral, and left–right. It also applies to other key features, including segments, appendages, and eyes. Unity underlies diversity.

There have been other key findings at the level of developmental genes. Among these are the importance of evolutionary changes in the regulatory sequences of genes and the importance of gene co-option – the evolutionary recruitment of a gene that initially performed one developmental role to take on another. These two things are linked with each other and with a third – the importance of 'replication and diversification' in evolution. More often referred to by its special-case name of 'duplication and divergence', this process has been at work over the aeons at various levels of biological organization, from genes to large-scale anatomical structures. When you have duplicate or replicate copies of something – be it a gene or a leg – natural selection has more room to manoeuvre. As long as one copy still fulfils its original role, another copy can be modified so that it fulfils a new one. In this way, homeobox genes have diversified into different families, and birds are able to soar through the air and yet still land on their feet.

Evo-devo is characterized in part by its discoveries. However, it is also characterized by the way it has shifted the focus of attention of evolutionary studies so that the individual animal or plant is once more a centre of attention, as it was in Darwin's day. In the intervening years, and especially during the half-century from about 1930 to 1980, the organism was displaced from its rightful place at the centre of things. For example, during that period we had a term that referred to all changes in genes – mutation – but no comparable term to describe all changes in developmental trajectories. That situation allowed authors of the time to focus on only one type of evolutionary change in development, to the virtual exclusion of others. Often, the focus was on heterochrony – changes in developmental timing. Heterochrony is important for sure, but its role should not be overstated. Other kinds of evolutionary change in development are also important.

Now, we have a term for all evolutionary changes in development – repatterning. And logically there are four subcategories of this, of which heterochrony is but one. The four types of developmental repatterning are as follows: changes in time (heterochrony), changes in space (heterotopy), changes in amount (heterometry), and changes in type (heterotypy). All are important; all have happened repeatedly in evolution and at different levels of organization. When we think of these four processes at the level of a developmental gene, we can see an important connection. Evolutionary changes in the regulatory regions of a gene are likely to lead to alterations in its place, time, or rate of expression, whereas changes in its coding region can lead to it making a different version of its protein – heterotypy. One interesting hypothesis is that the evolution of development is largely a matter of the first three of these processes rather than the last one.

I'd like to re-emphasize what I see as two of the main challenges for evo-devo in the future. The first is to understand the way in which biases in the production of variant developmental trajectories (developmental bias or channelling) interact with biases in their survival and reproduction (natural selection). From the limited work that has been conducted on this issue to date, it seems that selection can sometimes 'break' developmental bias, but sometimes not. The relative frequencies of these two outcomes of their interaction are of considerable interest. There is some way to go before we make headway on this issue. However, in the era of evo-devo we at least have a more balanced way of looking at things than before. Rather than thinking of selection as a positive force for change and developmental bias as an entirely negative effect – a hindrance to, or constraint on, selection – we can think of both selection and bias as having a dual positive–negative nature. Of course, the negative role of selection (the so-called purifying selection that removes maladapted variants) has been accepted for a long time. But the positive role of developmental bias has not.

The second main challenge for evo-devo is to achieve a much better understanding than we have at present of how evolutionary novelties (like the turtle's shell and the bird's beak) and body plans (like the vertebrate endoskeleton and the arthropod exoskeleton) arise. The key question here is whether such occasional origins are in some way different from the routine evolutionary changes in body size and shape that are happening all the time, and if so in what way they are different. I have advanced the hypothesis that the main difference lies in the nature of the available developmental variation rather than the nature of the selection that acts on the variation. However, not only is this just a hypothesis, but even if it is true we are still a long way from being able to generalize about the nature of the variants concerned. This issue is worth pursuing, despite its difficulty. It is all very well having case studies that can describe in detail how the sizes and shapes of bones change in the evolution of horses or humans. But if we can't explain how those bones evolved in the first place from an animal that lacked them, our science is that much the poorer. Evo-devo hasn't yet produced a satisfactory answer to such questions, but I suspect that one day it will.

Summary of Common Misunderstandings

Here, I list ten important issues where I think that there is a significant risk of misunderstandings among those who are new to the field. After each potential misunderstanding, there is a statement of the correct situation, as I perceive it. Many of these issues are related to the rationale underlying the emergence of evo-devo as a (relatively) new discipline.

When reading about these ten issues, a good approach is to ask yourself the question: do I agree that this is really a clear-cut case of misunderstanding versus correct understanding, or might it instead be a case of view *A* versus equally valid but contradictory view *B*? If you do this before reading the rest of the book, then I suggest that you do it again afterwards, and see to what extent your views have changed.

Evolution can be adequately described by a theory that is framed in terms of changes of gene frequencies in populations. The biological world spans several levels of organization, from the molecular, to the cellular and the organismic, up to the ecological. The approach to understanding evolution that is taken by population genetics focuses on genes (molecular level) and populations (ecological level). Understanding factors that influence the spread of new versions of genes in natural populations is *necessary* for a complete understanding of evolutionary processes. However, it is not *sufficient*. We also need to understand how organisms are built by developmental processes, and how these processes evolve.

The organism can be thought of as a suite of phenotypic characteristics. As we bring the organism back into evolutionary theory, we need to be careful how we do it. One common way of thinking about organisms is to use the dichotomy of genotype versus phenotype, or, to put it another way, the genes versus all the characteristics that they directly or indirectly make, ranging from proteins and membranes to feathers and eyes. This approach is flawed, however. The word 'phenotype' is three-dimensional shorthand for a four-dimensional developmental process. We can talk of the left-handed phenotype of a sinistrally coiled snail shell. However, what we are really trying to understand is a time-extended developmental process, the first part of which is a particular pattern of cell division in a shell-less early-stage embryo.

Developmental processes are completely determined by genes. Having decided to focus on developmental processes and their evolution, we should acknowledge that a developmental process is not simply a fixed trajectory of embryonic and juvenile body form through time, unwaveringly delivered as a result of having a particular set of genes. Rather, a developmental process is a combined product of genes, proteins, other molecular players within an organism, and influences coming into the organism from the outside – i.e. the environment. An example of the last of these is the determination of the sex of a turtle by the temperature at which it develops. Also, inasmuch as genes are involved in influencing developmental trajectories, they are not always those of the organism itself. In the case of right- versus left-handed snail shells, the direction of coiling of an individual snail is due not to its own genes but to those of its mother.

Evolution and development are independent processes operating at very different timescales. While the second part of the above assertion is broadly correct, the first part is definitely not. Within a particular evolutionary lineage at a particular time, the starting point for further evolutionary change in body form is the prevailing developmental trajectory, parts of which tend to be resistant to change and thus leave their mark on descendant creatures – as in the case of the gill clefts that human embryos possess because we are descended from fishy ancestors.

The role of development in evolution is always negative – as might be assumed from the common use of the phrase 'developmental constraint'. The developmental trajectory that characterizes a lineage at a particular point in evolutionary time channels subsequent evolutionary changes. This channelling can be both negative (some new trajectories being rendered improbable) and positive (some new trajectories being rendered probable). For this reason, a term that is now used more often than constraint is developmental bias. Natural selection can also be thought of as a bias – in the survival and reproductive probabilities of different genetic variants. I take the view that evolution is driven by the interaction between these two biases.

Natural selection always produces environment-specific adaptation. While all adaptation is driven by natural selection, not all selection, and not even all directional selection, produces adaptation in the sense of an environment-specific advantage. For example, while an evolutionary change in the size and shape of a bird's beak may be environment-specific, an evolutionary change that enhances *co*adaptation, for example better articulation of a newly evolved pair of jaws, may not be specific in this way; in other words it may apply (almost) regardless of the environment. Coadaptational changes are of particular interest in evo-devo.

Terms like evolutionary novelty and body plan are either precise and clear-cut on the one hand or so vague as to be useless on the other. The truth lies somewhere in between these two extreme views. As in other branches of biology, terms are often fuzzy. However, they are useful despite this. For example, neither the evolutionary origin of the vertebrate skeleton (a body-plan feature) nor the evolutionary origin of a novel type of shell in one vertebrate lineage (leading to turtles) is similar to the 'routine' evolution of characters such as body size, which is going on in all lineages all of the time.

Major changes in development, and hence body form, require alterations in many different genes. Two very similar-looking flies may have hundreds or even thousands of genetic differences between them. In contrast, two flies may differ radically in body form in that one of them may be 'normal' whereas the other has legs where its antennae should be as a result of a genetic difference in only a single gene or gene complex. The mapping between the genome and the pattern of development is not a simple one.

Sudden radical evolutionary change in the form of an adult animal is easy because some genes have key developmental roles. The part of the above sentence that is incorrect is the first part (before 'because'). Although what follows is true enough – some genes do indeed have major developmental roles – sudden radical evolutionary change requires the mutations concerned to be selectively advantageous. All the evidence that we have to date from both field and laboratory studies suggests that major-effect mutations are nearly always disadvantageous (or perhaps *always*, depending on how 'major' is defined).

Mutations contributing to evolutionary change can be arranged on a continuum from small to large effects on development. Despite the previous point – that mutations with major developmental effects are generally detrimental in terms of fitness – the use of a single spectrum of 'magnitude of developmental effect' is not an ideal way to classify mutations. Instead, we need to progress from that rather narrow perspective to a broader one in which we think of *kinds* of mutational effect on development. It now seems likely that the interesting evolutionary changes that we describe as the origins of novelties or body plans are neither (a) similar in all respects to more routine evolutionary changes nor (b) explicable in terms of individual revolutionary 'macromutations'. One of the main aims of evo-devo is to understand the different kinds of mutational effect on development that underlie interesting evolutionary transitions; and to consider these not as fixed changes in development but as genetic effects that interact with the rest of the genome and environmental factors in multifaceted ways.

References

Chapter 1

Bateson, W. 1894. *Materials for the Study of Variation, Treated with Especial Regard to Discontinuity in the Origin of Species*. Macmillan, London.

Darwin, C. 1859. *On the Origin of Species by Means of Natural Selection, or the Preservation of Favoured Races in the Struggle for Life*. John Murray, London.

Gilbert, S. F. and Epel, D. 2015. *Ecological Developmental Biology: The Environmental Regulation of Development, Health, and Evolution*, 2nd edition. Sinauer, Sunderland, MA.

Goodwin, B. 1994. *How the Leopard Changed its Spots: The Evolution of Complexity*. Weidenfeld & Nicolson, London.

Gould, S. J. 1977. *Ontogeny and Phylogeny*. Harvard University Press, Cambridge, MA.

Gould, S. J. and Lewontin, R. C. 1979. The spandrels of San Marco and the Panglossian paradigm: a critique of the adaptationist programme. *Proceedings of the Royal Society of London B*, 205: 581–598.

Kimura, M. 1983. *The Neutral Theory of Molecular Evolution*. Cambridge University Press, Cambridge.

Lewis, E. B. 1978. A gene complex controlling segmentation in *Drosophila*. *Nature*, 276: 565–570.

Lewontin, R. C. 1974. *The Genetic Basis of Evolutionary Change*. Columbia University Press, New York.

McGinnis, W., Garber, R. L., Wirz, J., Kuroiwa, A. and Gehring, W. J. 1984. A homologous protein-coding sequence in *Drosophila* homeotic genes and its conservation in other metazoans. *Cell*, 37: 403–408.

Moczek, A. P. (ed.) 2020. Special issue: Developmental bias in evolution. *Evolution & Development*, 22: 1–217.

Nüsslein-Volhard, C. and Wieschaus, E. 1980. Mutations affecting segment number and polarity in *Drosophila*. *Nature*, 287: 795–801.

Scott, M. P. and Weiner, A. J. 1984. Structural relationships among genes that control development: sequence homology between the *Antennapedia*, *Ultrabithorax* and *fushi tarazu* loci of *Drosophila*. *Proceedings of the National Academy of Sciences of the USA*, 81: 4115–4119.

Vedel, V., Chipman, A.D., Akam, M. and Arthur, W. 2008. Temperature-dependent plasticity of segment number in an arthropod species: the centipede *Strigamia maritima*. *Evolution & Development*, 10: 487–492.

Waddington, C. H. 1975. *The Evolution of an Evolutionist*. Edinburgh University Press, Edinburgh.

Williams, G. C. 1992. *Natural Selection: Domains, Levels, and Challenges*. Oxford University Press, New York.

Chapter 2

Bateson, W. 1894. *Materials for the Study of Variation, Treated with Especial Regard to Discontinuity in the Origin of Species*. Macmillan, London.

Clack, J. 2002. *Gaining Ground: The Origin and Early Evolution of Tetrapods*. Indiana University Press, Bloomington, IN.

Darwin, C. 1871. *The Descent of Man, and Selection in Relation to Sex*. John Murray, London.

De Beer, G. R. 1940. *Embryos and Ancestors*. Clarendon Press, Oxford.

Fisher, R. A. 1918. The correlations between relatives on the supposition of Mendelian inheritance. *Transactions of the Royal Society of Edinburgh*, 52: 399–433.

Fisher, R. A. 1930. *The Genetical Theory of Natural Selection*. Clarendon Press, Oxford.

Goldschmidt, R. 1940. *The Material Basis of Evolution*. Yale University Press, New Haven, CT.

Haeckel, E. 1866. *Generelle Morphologie der Organismen*. Georg Reimer, Berlin.

Haeckel, E. 1876. *The History of Creation: The Development of the Earth and its Inhabitants by the Action of Natural Causes*, in two volumes. Appleton, New York. (Various recent reprint editions are available.)

Haeckel, E. 1896. *The Evolution of Man: A Popular Exposition of the Principal Points of Human Ontogeny and Phylogeny*, in two volumes. Appleton, New York.

Haldane, J. B. S. 1932. *The Causes of Evolution*. Longman, London.

Panchen, A. L. 1992. *Classification, Evolution and the Nature of Biology*. Cambridge University Press, Cambridge.

Raff, R. A. 1996. *The Shape of Life: Genes, Development and the Evolution of Animal Form*. Chicago University Press, Chicago.

Raff, R. A. and Kaufman, T. C. 1983. *Embryos, Genes, and Evolution: The Developmental Genetic Basis of Evolutionary Change*. Macmillan, New York.

Richards, R. J. 2008. *The Tragic Sense of Life: Ernst Haeckel and the Struggle over Evolutionary Thought*. Chicago University Press, Chicago.

Santayana, G. 1905. *The Life of Reason: The Phases of Human Progress*, in five volumes. Charles Scribner's Sons, New York.

Thompson, D'A. W. 1917. *On Growth and Form*. Cambridge University Press, Cambridge.

Von Baer, K. E. 1828. *Uber Entwicklungsgeschichte der Tiere: Beobachtung und Reflexion*. Borntrager, Königsberg.

Chapter 3

Aguinaldo, A. M. A. *et al.* (7 authors) 1997. Evidence for a clade of nematodes, arthropods and other moulting animals. *Nature*, 387: 489–493.

Berrill, N. 1961. *Growth, Development, and Pattern*. Freeman, San Francisco, CA.

Brown, W. L. 1958. General adaptation and evolution. *Systematic Zoology*, 7: 157–168.

Darwin, C. and Wallace, A. R. 1858. On the tendency of species to form varieties; and on the perpetuation of varieties and species by natural means of selection. *Zoological Journal of the Linnean Society*, 3: 46–62.

Dawkins, R. 1986. *The Blind Watchmaker*. Longman, London.

Huxley, J. S. 1942. *Evolution: The Modern Synthesis*. Allen and Unwin, London.

Minelli, A. 2009. *Perspectives in Animal Phylogeny and Evolution*. Oxford University Press, Oxford.

Minelli, A. 2021. *Understanding Development*. Cambridge University Press, Cambridge.

Chapter 4

Arthur, W. 1997. *The Origin of Animal Body Plans: A Study in Evolutionary Developmental Biology*. Cambridge University Press, Cambridge.

Arthur, W. 2011. *Evolution: A Developmental Approach*. Wiley-Blackwell, Oxford.

Carroll, S. B., Grenier, J. K. and Weatherbee, S. D. 2005. *From DNA to Diversity: Molecular Genetics and the Evolution of Animal Design*, 2nd edition. Blackwell, Oxford.

Duboule, D. 1994. Temporal colinearity and the phylotypic progression: a basis for the stability of a vertebrate Bauplan and the evolution of morphologies through heterochrony. *Development (Supplement)*, 1994:135–142.

Gould, S. J. and Lewontin, R. C. 1979. The spandrels of San Marco and the Panglossian paradigm: a critique of the adaptationist programme. *Proceedings of the Royal Society of London B*, 205: 581–598.

Kirschner, J. and Gerhart, J. 1998. Evolvability. *Proceedings of the National Academy of Sciences of the USA*, 95: 8420–8427.

Klingenberg, C. K. 2010. Evolution and development of shape: integrating quantitative approaches. *Nature Reviews Genetics*, 11: 623–635.

Roth, G. and Wake, D. B. 1985. Trends in the functional morphology and sensorimotor control of feeding behaviour in salamanders: an example of the role of internal dynamics in evolution. *Acta Biotheoretica*, 34: 175–192.

Wilson, E. O. and Bossert, W. H. 1971. *A Primer of Population Biology*. Sinauer/ Oxford University Press, Oxford.

Chapter 5

Allen, C. E., Beldade, P., Zwann, B. J. and Brakefield, P. 2008. Differences in the selection response of serially repeated color pattern characters: standing variation, development, and evolution. *BMC Evolutionary Biology*, 8: 94 (13 pages).

Beldade, P., Koops, K. and Brakefield, P.M. 2002. Developmental constraints versus flexibility in morphological evolution. *Nature*, 416: 844–847.

Böhmer, C., Amson, E., Arnold, P., van Heteren, A.H. and Nyakatura, J.A. 2018. Homeotic transformations reflect departure from the mammalian 'rule of seven' cervical vertebrae in sloths: inferences on the *Hox* code and morphological modularity of the mammalian neck. *BMC Evolutionary Biology*, 18: 84 (11 pages).

Chipman, A. D., Arthur, W. and Akam, M. 2004. A double segment periodicity underlies segment generation in centipede development. *Current Biology*, 14: 1250–1255.

Galis, F. 1999. Why do almost all mammals have seven cervical vertebrae? Developmental constraints, *Hox* genes, and cancer. *Journal of Experimental Zoology*, 285: 19–26.

Gould, S. J. 1983. *Hen's Teeth and Horse's Toes: Further Reflections in Natural History*. Norton, New York.

Gould, S. J. and Lewontin, R. C. 1979. The spandrels of San Marco and the Panglossian paradigm: a critique of the adaptationist programme. *Proceedings of the Royal Society of London B*, 205: 581–598.

Haldane, J. B. S. 1932. *The Causes of Evolution*. Longman, London.

Kemp, T. S. 2016. *The Origin of Higher Taxa: Palaeobiological, Developmental, and Ecological Perspectives*. Oxford University Press, Oxford.

Simpson, G. G. 1944. *Tempo and Mode in Evolution*. Columbia University Press, New York.

Varela-Lasheras, I., Bakker, A. J., van der Mije, S. D., Metz, J. A. J., van Alphen, J. and Galis, F. 2011. Breaking evolutionary and pleiotropic constraints in mammals: on sloths, manatees and homeotic mutations. *EvoDevo*, 2: 11 (27 pages).

Chapter 6

Dawkins, R. 1986. *The Blind Watchmaker*. Longman, London.

Dugon, M. M., Hayden, L., Black, A. and Arthur, W. 2012. Development of the venom ducts in the centipede *Scolopendra*: an example of recapitulation. *Evolution & Development*, 14: 515–521.

Grande, C. and Patel, N. 2009. Nodal signalling is involved in left–right asymmetry in snails. *Nature*, 457: 1007–1011.

Haldane, J. B. S. 1932. *The Causes of Evolution*. Longman, London.

Held, L. I. 2014. *How the Snake Lost its Legs: Curious Tales from the Frontier of Evo-Devo*. Cambridge University Press, Cambridge.

Hughes, C. L. and Kaufman, T. C. 2002. Hox genes and the evolution of the arthropod body plan. *Evolution & Development*, 4: 459–499.

Johnson, M. S. 1982. Polymorphism for direction of coil in *Partula suturalis*: behavioural isolation and positive frequency dependent selection. *Heredity*, 49: 145–151.

Loredo, G.A. *et al.* (12 authors) 2001. Development of an evolutionarily novel structure: fibroblast growth factor expression in the carapacial ridge of turtle embryos. *Journal of Experimental Zoology*, 291: 274–281.

Ohno, S. 1970. *Evolution by Gene Duplication*. Springer-Verlag, New York.

Wagner, G. P. 2014. *Homology, Genes, and Evolutionary Innovation*. Princeton University Press, Princeton, NJ.

Chapter 7

Akam, M. 1998. Hox genes: from master genes to micromanagers. *Current Biology* 8: R676–R678.

Arthur, W. 1984. *Mechanisms of Morphological Evolution: A Combined Genetic, Developmental and Ecological Approach*. Wiley, Chichester.

Clark, R. B. 1964. *Dynamics in Metazoan Evolution: The Origin of the Coelom and Segments*. Clarendon Press, Oxford.

Cuvier, G. 1817. *Le Règne Animal Distribué d'après son Organisation, pour servir de base a l'histoire naturelle des animaux et d'introduction a l'anatomie comparée*, 1st edition, in four volumes. Deterville, Paris.

Geoffroy Saint-Hilaire, E. 1822. Considérations générales sur la vertèbre. *Mémoires du Museum national d'Histoire naturelle*, 9: 89–119.

Haldane, J. B. S. 1932. *The Causes of Evolution*. Longman, London.

Lewis, E. B. 1978. A gene complex controlling segmentation in *Drosophila*. *Nature*, 276: 565–570.

Riedl, R. 1978. *Order in Living Organisms: A Systems Analysis of Evolution*. Wiley, Chichester.

Salser, S. J. and Kenyon, C. 1996. A *C. elegans* Hox gene switches on, off, on and off again to regulate proliferation, differentiation and morphogenesis. *Development* 122: 1651–1661.

Thomson, K. 1988. *Morphogenesis and Evolution*. Oxford University Press, Oxford.

Chapter 8

Balavoine, B. and Adoutte, A. 2003. The segmented *Urbilateria*: a testable scenario. *Integrative and Comparative Biology*, 43: 137–147.

Chipman, A. D. 2010. Parallel evolution of segmentation by co-option of ancestral gene regulatory networks. *BioEssays*, 32: 60–70.

Gilbert, S. 2016. *Developmental Biology*, 11th edition. Sinauer, Sunderland, MA.

Hejnol, A. and Martindale, M. Q. 2008. Acoel development supports a simple planula-like urbilaterian. *Philosophical Transactions of the Royal Society of London B*, 363: 1493–1501.

Jiggins, C. D., Wallbank, R. W. R. and Hanly, J. J. 2017. Waiting in the wings: what can we learn about gene co-option from the diversification of butterfly wing patterns? *Philosophical Transactions of the Royal Society of London B*, 372: 20150485 (10 pages).

Kozmik, Z. 2005. Pax genes in eye development and evolution. *Current Opinion in Genetics and Development*, 15: 430–438.

Panganiban, G. S. *et al.* (14 authors) 1997. The origin and evolution of animal appendages. *Proceedings of the National Academy of Sciences of the USA*, 94: 5162–5166.

Smith, J. L. B. 1939. A living fish of Mesozoic type. *Nature*, 143: 455–456.

True, J. R. and Carroll, S. B. 2002. Gene co-option in physiological and morphological evolution. *Annual Review of Cell and Developmental Biology*, 18: 53–80.

Watson, J. D. and Crick, F. H. C. 1953. Molecular structure of nucleic acids: a structure for deoxyribose nucleic acid. *Nature*, 171: 737–738.

Chapter 9

Carroll, R. 2000. Towards a new evolutionary synthesis. *Trends in Ecology and Evolution*, 15: 27–32.

Carroll, S. B. 2008. Evo-devo and an expanding evolutionary synthesis: a genetic theory of morphological evolution. *Cell*, 134: 25–36.

Darwin, C. 1859. *On the Origin of Species by Means of Natural Selection, or the Preservation of Favoured Races in the Struggle for Life*. John Murray, London.

De Beer, G. R. 1940. *Embryos and Ancestors*. Clarendon Press, Oxford.

Fisher, R. A. 1930. *The Genetical Theory of Natural Selection*. Clarendon Press, Oxford.

Gould, S. J. 1977. *Ontogeny and Phylogeny*. Harvard University Press, Cambridge, MA.

Gould, S. J. 2002. *The Structure of Evolutionary Theory*. Harvard University Press, Cambridge, MA.

Haldane, J. B. S. 1932. *The Causes of Evolution*. Longman, London.

Huxley, J. S. 1942. *Evolution: The Modern Synthesis*. Allen & Unwin, London.

Laland, K. *et al.* 2014. Does evolutionary theory need a rethink? Yes, urgently. *Nature*, 514: 161–164. (One of a pair of linked articles; see also Wray, Hoekstra *et al.*)

Pigliucci, M. and Müller, G. (eds) 2010. *Evolution: The Extended Synthesis.* MIT Press, Cambridge, MA.

Waddington, C. H. 1957. *The Strategy of the Genes.* Allen & Unwin, London.

Wray, G., Hoekstra, H. *et al.* 2014. Does evolutionary theory need a rethink? No, all is well. *Nature*, 514: 161–164. (One of a pair of linked articles; see also Laland *et al.*)

Wright, S. 1931. Evolution in Mendelian populations. *Genetics*, 16: 97–159.

Yampolsky, L.Y. and Stoltzfus, A. 2001. Bias in the introduction of variation as an orienting factor in evolution. *Evolution & Development*, 3: 73–83.

Figure Credits

Most of the figures are originals, produced specifically for this book. These were all done by my son, Stephen Arthur, whom I thank for his patience in using my rudimentary scribblings as starting points for his far more accomplished artwork. He also redrew the other figures from various sources, as follows: Figures 2.1 and 7.3, redrawn from Arthur, *The Origin of Animal Body Plans*, Cambridge University Press, 1997; Figure 2.2, redrawn from Thompson, *On Growth and Form*, Cambridge University Press, 1942; Figure 4.1, redrawn from Arthur, *Evolving Animals*, Cambridge University Press, 2014; Figures 5.2 and 5.3, redrawn from Arthur, *Evolution: A Developmental Approach*, Wiley-Blackwell, 2011.

Index